# Springer Advanced Texts in Chemistry

Charles R. Cantor, Editor

# Springer Advanced Texts in Chemistry

*Series Editor:* Charles R. Cantor

Robert K. Scopes

# Protein Purification

## Principles and Practice

With 145 Figures

Springer-Verlag
New York   Heidelberg   Berlin

Robert K. Scopes
Department of Biochemistry
La Trobe University
Bundoora, Victoria 3083
Australia

*Series Editor:*
Charles R. Cantor
Columbia University
College of Physicians and Surgeons
Department of Human Genetics and Development
New York, New York 10032, U.S.A.

Sponsoring Editor: Philip Manor
Production: Kate Ormston


*Cover:* Protein (green) with charges and hydrophobic areas (see p. 40)
against a background of polyacrylamide (dark brown).


Library of Congress Cataloging in Publication Data
Scopes, R. K. (Robert K.)
    Protein purification.
    (Springer advanced texts in chemistry)
    Bibliography: p.
    Includes index.
    1. Enzymes—Purification.   I. Title.   II. Series.
QP601.S39   1982      547.7'58      82-10476

Typeset by University Graphics, Inc., Atlantic Highlands, NJ.
Printed and bound by R. R. Donnelley & Sons, Harrisonburg, VA.
Printed in the United States of America.

9 8 7 6 5 4 3 2   (Second Printing, 1984)

ISBN 0-387-90726-2   Springer-Verlag New York Heidelberg Berlin
ISBN 3-540-90726-2   Springer-Verlag Berlin Heidelberg New York

# Series Preface

New textbooks at all levels of chemistry appear with great regularity. Some fields like basic biochemistry, organic reaction mechanisms, and chemical thermodynamics are well represented by many excellent texts, and new or revised editions are published sufficiently often to keep up with progress in research. However, some areas of chemistry, especially many of those taught at the graduate level, suffer from a real lack of up-to-date textbooks. The most serious needs occur in fields that are rapidly changing. Textbooks in these subjects usually have to be written by scientists actually involved in the research which is advancing the field. It is not often easy to persuade such individuals to set time aside to help spread the knowledge they have accumulated. Our goal, in this series, is to pinpoint areas of chemistry where recent progress has outpaced what is covered in any available textbooks, and then seek out and persuade experts in these fields to produce relatively concise but instructive introductions to their fields. These should serve the needs of one semester or one quarter graduate courses in chemistry and biochemistry. In some cases the availability of texts in active research areas should help stimulate the creation of new courses.

New York                                                CHARLES R. CANTOR

# Preface

This book marks twenty years of research involving enzyme purifications, initiated partly because at the time of my higher degree (I needed some creatine kinase for ATP regeneration) we could not afford to buy commercial purified enzymes; there were hardly any available at that time anyway. The fascination of isolating a reasonably pure enzyme from a complex natural soup of proteins remains with me; as a challenge and an academic exercise I still spend much time on enzyme purification even when I have no real use for the final product! In those days everything was empirical: try an ammonium sulfate cut; try organic solvents; will it adsorb on calcium phosphate gel? There was not much else available. Now there is such a vast range of methods, of materials, and of approaches to the problem of enzyme purification that the difficulties lie not in the subtleties of manipulating a few available techniques, but in trying to decide which of the plethora of possibilities are most suitable.

It is the purpose of this book to guide the newcomer through the range of protein fractionation methods, while pointing out the advantages and disadvantages of each, so that a choice can be made to suit the problem at hand. Thus, traditional procedures such as salt fractionation are presented, along with the many modern developments in affinity chromatography and related techniques, which have been so successful in many cases, but have not proved to be the answer to all problems. During the two years of preparation of the manuscript, many new techniques and, equally important, new commercial products have been reported; preliminary drafts were updated as a new product came on the market, but by the time this is published there will undoubtedly be still more. In some cases the new product will be so superior that it will quickly supersede the older variety. Increasingly often, however, the question of cost-effectiveness is raised: this new system may be the best, but will it ruin our budget? For this reason I feel that the older, cheaper products (and more classical techniques) will be with us for many years to come.

A few sentences on what this book does, and does not attempt to do: It is not a comprehensive presentation of all the methods ever used in enzyme purification, nor does it give detailed examples or precise instructions; if it did, it would occupy far more shelf space and would only duplicate excellent multivolume treatises already available—in particular the series *Methods in Enzymology* and *Laboratory Techniques in Biochemistry and Molecular Biology,* and *The Proteins* (3rd edn., Vol. 1, eds. Newarth and Hill). Instead, it gives a brief account of the main procedures available, with some simple (even simplistic) theoretical and thermodynamic explanations of the events occurring. A basic background in biochemistry and protein chemistry is assumed: I expect the reader already to have on his shelves textbooks describing protein structure, simple enzyme kinetics, and thermodynamics. It is aimed to assist all students and researchers involved in the process of isolating an enzyme, from whatever source, whether it be a new project or simply following a published procedure.

In most places in this book the words "enzyme" and "protein" can be interchanged. All enzymes are proteins, but the reverse is not so. For the most part, I have adopted the system of using "enzyme" when referring to the particular protein being purified (even though it may not be an enzyme), and "protein" when referring either to the complete mixture at hand, or else to the proteins *other than* the particular one being purified.

There are many audiences to satisfy in a book such as this. Three, not necessarily exclusive, categories are: (i) those purifying a protein that has never successfully been purified before from any source; (ii) those purifying a protein from a new source, there being an adequate method available using some other source; and (iii) those who are simply following a recipe in an attempt to obtain a pure protein equivalent to that reported previously. As to the first category: it is becoming less common now for anyone to be purifying an enzyme for the first time, simply because there are fewer enzymes being discovered and, one presumes, few to be discovered. But in the second category, it is more common than it used to be for people to be purifying enzymes from new sources: no longer do biochemists restrict themselves to *E. coli,* yeast, rat liver, rabbit muscle, pig heart and spinach! Even a small shift in evolutionary terms (e.g., from pig to bovine heart) results in different behaviors not only of the enzyme being isolated but, equally importantly, of the other proteins present. Many standard methods are so critically dependent on precise conditions that a shift in source material may result in complete failure because of a minor variation in protein properties. Nevertheless, it is rare for the one enzyme to vary much in molecular weight even over large evolutionary distances; within phyla other properties such as ion exchange behaviors (dependent on isoelectric points) are not often widely at variance, and solubility behavior may well be similar.

The third category contains those who wish only to duplicate a preparation that someone else has reported. If sufficient important details have been published, and the raw material and laboratory operating conditions are essentially identical, then there is no need for further help; unfortunately these criteria are not always met. Without experience and general knowledge of the principles, most beginners are unable to reproduce a method the first time—or even at all. Also, if it does not work they are afraid to depart one iota from the written word in case things get worse. For these people my advice will be to make use of the information given but, if things are not going well, not to be afraid to change the conditions (they may have been reported incorrectly anyway, and a misprint not spotted in proof-reading could waste months); also, not to worry if they do not have a PX28 rotor for an SS26 model D centrifuge, if all that is needed is the pellet. If things are still going wrong, then it is time to introduce greater variations, using new materials or methods that have been developed since the original publication. Because enzyme purification is essentially a methodology-oriented practice, it is usually worth manipulating conditions repeatedly until the ideal combinations are found. This can be time-consuming, but is worth the effort in the end, as you can be sure of reproducibility if you are confident of the limitations on each step.

Finally I wish to acknowledge the contributions over the years of many students and the staff at La Trobe University. I should also like to thank Prof. K. Mosbach, Dr. C. R. Lowe, Dr. I. P. Trayer, Dr. C.-Y. Lee and Dr. F. von der Haar for making available reprints and preprints of their work on enzyme purification. Financial support from the Australian Research Grants Committee is gratefully acknowledged.

Bundoora, Victoria           ROBERT K. SCOPES
Australia

# Contents

Chapter 4

**Separation by Adsorption**                                           67

Chapter 5

**Separation in Solution**                                             151

Chapter 6

**Maintenance of Active Enzymes**                                      185

Chapter 7

**Optimization of Procedures and Following a Recipe**                  201

Chapter 8

**Measurement of Enzyme Activity**                                     213

Chapter 9

**Analysis for Purity; Crystallization**                                                          **245**

Appendix A                                                                                                **261**

Appendix B
**Solutions for Measuring Protein Concentration**                                    **265**

# Chapter 1
# The Enzyme Purification Laboratory

This chapter introduces the basic methods and equipment used in enzyme purification. It is entirely possible to purify enzymes using relatively simple equipment; the bare essentials might be considered a centrifuge and a UV spectrophotometer. Of course, access to a wider range of equipment and automated systems can make life much easier. But more elaborate equipment and procedures have their cost. It will be stressed many times in this book that *delays during isolation of any enzyme must be avoided*. Any procedure that relies on using equipment which is slow, unreliable, or sometimes unavailable is dangerous. By relying only on relatively simple and common apparatus, delays can usually be avoided. The message is clear: always use the simplest of procedures first, and use specialized methods only when all else has failed.

## 1.1 Apparatus, Special Materials, and Reagents

There are many complex, sophisticated pieces of equipment designed for protein separations, and as in every other walk of life, these are becoming more automated for convenience and simplicity of operation. Yet as the apparatus becomes more enclosed in black boxes, controlled by microprocessors, we become further removed from the realities of what is happening, and in some ways have less control over what we want to do. Most of the more complex equipment is designed for repeating routine operations reliably, and as such is not always the most appropriate for developing new methods. But apart from these considerations, there is the one of cost effectiveness; duplication of cheap equipment is often more useful than getting one of the most expensive on the market; money spent on the simplest things, such as plenty of pipettes, test tubes, and beakers, can be a better investment than purchasing an esoteric apparatus for carrying out one particular type of protein separation process that may not be used often (and may break down).

Having said that, there remains a baseline of minimal apparatus which any enzyme purification laboratory must have available. Measures of enzyme activity will usually be based on either spectrophotometry or radiochemistry. Although one can do without a scintillation counter if all assays can be done spectrophotometrically, a spectrophotometer is indispensable, since it is required for protein concentration measurements (cf. section 8.5). A spectrophotometer with UV lamp, preferably with an attached recorder for timed reactions, is regarded as a prime requirement.

Centrifuging can rarely be avoided. On an industrial scale filter systems are commonly used for separating precipitates, but in the laboratory a centrifuge is more convenient. The principles and practices of centrifuging are described below (cf. section 1.2). For most purposes the standard workhorse centrifuge is a refrigerated instrument capable of maintaining temperatures at or somewhat below $0°C$, while centrifuging ability, expressed as relative $g$ force $\times$ capacity in liters, should be of the order 10,000–15,000. Many other types of centrifuge can be useful, from the benchtop small-scale machine to preparative ultracentrifuges. But these are not generally necessary and would rarely get as much use in enzyme purification work as the basic machine described above.

Column chromatography is so generally employed that equipment for this is essential. Although column work can be carried out with a home-made glass tube or a syringe (cf. section 1.3), with manual fraction-collecting and subsequent spectrophotometric measurements on each tube, much more work can be done more easily with the basic automated setup, which costs about the same as the spectrophotometer and less than the centrifuge. This would consist of a variety of sizes of columns, a fraction collector with UV monitor, a peristaltic pump, and a magnetic stirrer (for gradient formation). A wide variety of column sizes and shapes should be available so that the optimum amount of column packing material can be used for the sample available; columns with adjustable plungers are advisable. A range of sizes suitable for "desalting" by gel filtration is described in section 1.4. For adsorption chromatography still other sizes may be required. A given amount of money may be better spent on two or more columns of simple design rather than one made to optimum specifications.

Homogenizers for disrupting cells are needed, and are described in section 2.2. Routine laboratory equipment such as balance, pH meter, stirrers, and ice machines are as necessary in enzyme purification as in other biochemical methodology. Volumetric measurements can be made with graduated cylinders (rarely is the accuracy of a volumetric flask necessary), pipettes, and, for small amounts, microliter syringes. Enzyme assays frequently demand the addition of a very small sample of the test solution, perhaps less than 1 $\mu$l. Modern syringes are capable of delivering such volumes with sufficient accuracy. A range of syringes from 0.5 up to 250 $\mu$l is highly desirable. In addition, continuously adjustable "Eppendorf"-type pipettes, using disposable tips, are useful for volume ranges from 5 $\mu$l up to 5 ml.

Analysis of protein mixtures by electrophoresis is carried out by gel electrophoresis or isoelectric focusing, most often on polyacrylamide gel which may or may not contain dodecyl sulfate as denaturant. Methodology is described in section 9.1. Many excellent equipment designs are now available commercially; starting from scratch a newcomer would be advised to obtain a thin-slab apparatus. There can be few biochemical laboratories in the world which do not have any equipment for analytical gel electrophoresis.

There are a number of special reagents that the protein purifier requires. These are mainly packing materials for column chromatography and chemicals that are frequently used. Ion exchangers (section 4.3), gel filtration materials (section 5.1), and affinity adsorbents (section 4.5) should be on hand. More mundane, but vitally important are such things as dialysis tubing (a range of sizes), a range of buffers (section 6.1), bulk ammonium sulfate of adequate purity (section 3.3), thiol components such as 2-mercaptoethanol and dithiothreitol, EDTA (ethylene diamine tetraacetic acid), sodium dodecyl sulfate, protamine sulfate, phenylmethylsulfonyl fluoride, Folin–Ciocalteau reagent, Coomassie blue, etc., the uses for which will be described in various places throughout this book.

# 1.2  Separation of Precipitates and Particulate Material

## Filtration

The development of efficient refrigerated centrifuges provided a cleaner, more efficient method than filtration for the separation of precipitated proteins and other matter, especially on a small scale. However, there are still occasions when filtration is preferable. On a very large scale, handling of large volumes by centrifugation may require nonstandard equipment such as continuous-flow adaptors (Figure 1.1), and in any case the process will take a long time, perhaps comparable with large-scale filtration. On the other hand, the size and softness of most biological precipitates lead to rapid clogging of filtration materials so that after an initial burst of clear filtrate, the flow reduces to a trickle, even under suction, and the filtration material must be replaced. On an industrial scale it is possible to use special rigid filters with automatic scraping equipment for removing the slimelike precipitate as it collects, but such systems are not normally available in a laboratory.

Filtration can be greatly improved using filter aids such as Celite, a diatomaceous earth consisting mainly of $SiO_2$ (see Figure 1.2) (as used for swimming pools). The best results are obtained by mixing in the Celite with the sample to be filtered, then pouring the suspension onto a Büchner funnel under slight suction. By creating a very large surface to trap the gelatinous precipi-

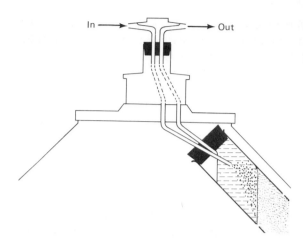

**Figure 1.1.** Simplified diagram of the continuous-flow adaptor for Sorvall SS-34 centrifuge rotor. Liquid can be passed through at several hundred milliliters per minute, collecting particulate matter up to 300 cm³ in volume.

tate, the filter aid allows much more filtrate to be obtained before eventually clogging up. The process is mainly used for removing small amounts of unwanted particulate material; it cannot deal with large quantities of precipitate successfully.

For removing obviously lumpy precipitates, hair, dirt, insoluble salt residues, etc., filtration through a no. 54 (Whatman) filter paper can be carried out; this is particularly appropriate immediately prior to a column procedure where the precipitate would otherwise be "filtered" on the packing material. Centrifugation is usually equally satisfactory.

A recent development in large-scale "ultrafiltration" (applying pressure to force liquid through a membrane while retaining particles too large to pass through it) is the Pellicon Cassette System from Millipore Corporation. Using a total filter area of up to 2 m², small particles are retained, removing filtrate at rates of up to 1 liter/min. This system is particularly useful for concentrating microorganisms, before centrifuging to obtain a proper precipitate; it does not itself produce a "paste," just a more concentrated suspension.

## Centrifugation

Centrifugation relies on the sedimentation of particles in an increased gravitational field. The forces acting on a particle are illustrated in Figure 1.3. For rigid spherical particles the time required for a particle to sediment in a given medium from the meniscus to the bottom of the tube is given by

$$t = \frac{9}{2} \frac{\eta}{\omega^2 r^2 (\rho - \rho_0)} \ln \frac{x_b}{x_t} \tag{1.1}$$

where $x_t$ = radial distance of meniscus
$x_b$ = radial distance of bottom of tube

and the other symbols are defined in the caption to Figure 1.3.

Thus the time required is proportional to the viscosity, but inversely proportional to the density difference and to the square of both the particle radius and angular velocity. The other factors are geometrical considerations for a particular rotor. From this we see that, other things being equal, a particle half the radius requires twice the angular velocity to sediment in the same time. More critical is the density difference; this is large for aqueous solutions containing little solute, but when there is a salt concentration such as in salting out (cf. section 3.3), this factor can be quite small. The density of anhydrous protein is about 1.34, and an aggregated protein particle contains about 50% protein and 50% trapped solution. Thus $\rho$ for an aggregate in water would be about $\frac{1}{2}(1.34 + 1.00) = 1.17$ ($\rho - \rho_0 = 0.17$), whereas in 80% saturated ammonium sulfate it would be close to 1.27, the solution $\rho_0$ being 1.20 ($\rho - \rho_0 = 0.07$). On the other hand, 33% acetone has a density of 0.93, for which $\rho - \rho_0$ works out as 0.20, even greater than for pure water. This illustrates one advantage of using organic solvent precipitation (cf. section 3.4), especially for initially dense, viscous solutions, i.e., the precipitates sediment more rapidly.

The optimum objective of centrifugation in enzyme purification is to obtain a tightly packed precipitate and a clear supernatant; exceeding the minimum centrifugation time for this to occur is no disadvantage. With precipitates formed by salting out, isoelectric precipitation, organic solvents, and most other precipitants described in Chapter 3, centrifugation at about "15,000 $g$" for 10

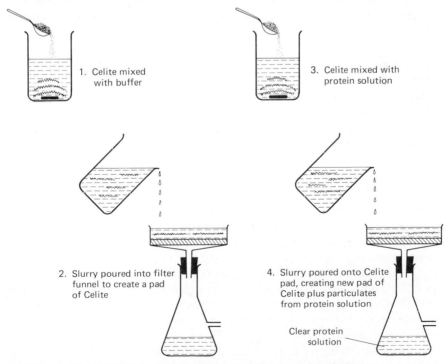

1. Celite mixed with buffer

3. Celite mixed with protein solution

2. Slurry poured into filter funnel to create a pad of Celite

4. Slurry poured onto Celite pad, creating new pad of Celite plus particulates from protein solution

Clear protein solution

**Figure 1.2.** Use of Celite as a filter aid for clarifying protein solutions.

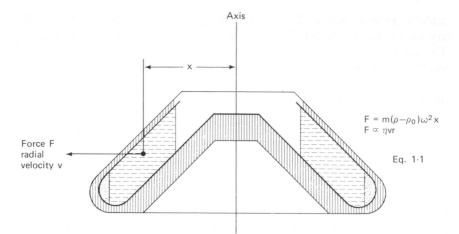

**Figure 1.3.** Forces acting on a particle in a centrifuge rotor. Mass of particle = $m$, density of particle = $\rho$, density of solution = $\rho_0$, angular velocity of rotor = $\omega$, viscosity of solution = $\eta$, and Stokes's radius of particle = $r$.

min, or "5000 $g$" for 30 min is usually sufficient. Note that on a small scale "5000 $g$" for 30 min might be adequate in a centrifuge tube with sedimentation distance of 5 cm, but on a large scale where the distance might be as much as 20 cm, 30 min may not be enough. The $g$ values above are average figures, since at a given rotation speed the $g$ value increases as the particle sediments. However if one quotes not $g$ values but rotation speed $\omega$, Eq. (1.1) indicates that it is not the absolute distance of sedimentation that determines the time taken, but the logarithm of the ratio between $x_b$ and $x_i$; this dimensional factor does not vary much between different centrifuges. Thus "5000 $rpm$ for 30 min" would produce a similar result in all centrifuges.

Particle size is clearly critical in determining the amount of centrifuging required. Any precipitation process will result in a range of particle sizes, but the numbers quoted above will generally be sufficient to sediment all aggregated protein that forms an observable cloudiness; smaller protein aggregates that are not sedimented are also not likely to be visible in the solution. But many cellular organelles, and fractions derived from them, are not by any means pure protein. They can contain much lipid material, which drastically reduces the term $(\rho - \rho_0)$ in Eq. (1.1)—in fact, this term may even become negative, in which case the particles move to the top of the tube as a floating layer.

Crude homogenates containing organelles or fragments can be fractionated by differential centrifugation processes in which the value of $\omega^2 t$ is increased at each step, causing particles with successively smaller $r^2 (\rho - \rho_0)$ values to be sedimented. Differential centrifugation carried out in this way is a first step in purifying organelles from a variety of tissues if the enzyme required is present not in the soluble fraction but in a particulate form.

Modern centrifuges are sturdy pieces of equipment, designed to survive a certain amount of mishandling, and to switch off automatically if things go very wrong. The main thing that can be wrong is an imbalance in the rotor. This may be caused by a tube cracking during the run, or by a misbalance of the tubes in the first place. Obviously tubes must be placed opposite each other or, in the case of heads with 6 spaces (or a greater multiple of 3), 3 equally balanced tubes arranged evenly (see Figure 1.4).

Although balancing by volume by eye is adequate for smaller tubes, volumes larger than 200 ml should be weighed to avoid imbalance. However, this refers to the situation when all tubes contain the same liquid. To balance a single tube containing a sample in, say, 80% saturated ammonium sulfate (density about 1.20) with water, one could allow for the extra weight by increasing the water volume (Figure 1.5). But this is not strictly correct since (a) it is the inertia, not the mass of the liquid that should be equal, and (b) as particles sediment in the sample tube the inertia increases. It is better to use two tubes less than half full (provided that the tubes are rigid enough to resist the $g$ forces) or to use smaller tubes in adaptors or in a different rotor. Centrifuge tubes made from polypropylene or polycarbonate are strong enough for most purposes. Glass centrifuge tubes may be desirable on occasions, but conventional glass (Pyrex) centrifuge tubes withstand only 3–4000 $g$; special toughened glass (Corex) tubes are available for forces up to 15,000 $g$.

Finally here are a few words on the care of centrifuges. Although biochemists rarely use strongly corrosive acids or alkalis, it is not always appreciated how corrosive mere salt can be. Centrifuge rotors cast from aluminium alloys corrode badly if salts, such as ammonium sulfate solution, are in contact with the metal for a period. Any spillage should be immediately rinsed away; it is best to rinse and clean a centrifuge rotor after every use, or at least every day.

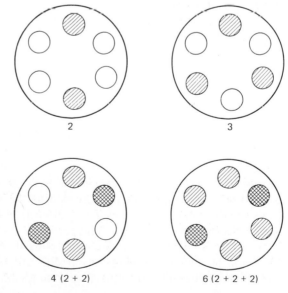

**Figure 1.4.** Arrangements of 2, 3, 4, and 6 tubes in a 6-place centrifuge rotor. In the cases of 4 and 6, only the tubes opposite each other need to be identically balanced (see shading).

2

3

4 (2 + 2)

6 (2 + 2 + 2)

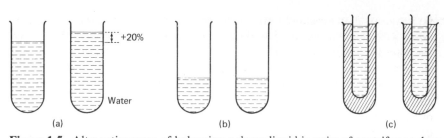

**Figure 1.5.** Alternative ways of balancing a dense liquid in pairs of centrifuge tubes: (a) On a small scale a rough balance is obtained with extra water to equalize masses. (b) Partial filling of both tubes with the dense liquid is preferable to (a), provided that the tubes are sufficiently rigid to avoid collapse when not filled. (c) The ideal situation is to have two tubes filled with the dense liquid, using adaptors to fit the rotor head, if necessary.

Contamination of the outside of centrifuge tubes with salt solutions (they may have been immersed in an ice-salt bath for cooling) can result in corroded pits inside the rotor if it is not washed frequently. It is also worth noting that constant use of the machine at top speed will wear out both rotor and motor more quickly; for many manipulations in enzyme purification half-speed is quite adequate.

## 1.3   Principles of Column Chromatography

Much of the detailed discussion of operating ion exchange columns, gel filtration, etc. will be described in Chapters 4 and 5. This short introduction gives the fundamental principles and describes the apparatus used in column chromatographic procedures.

Chromatography involves a retardation of solutes with respect to the solvent front progressing through the material. The name literally means "color drawing" and was used to describe the separation of natural pigments on filter papers by differential retardation (Figure 1.6). This same principle is still widely used—it can be called one-dimensional chromatography since there is no substantial thickness to the paper. A second dimension is introduced by having a column of matrix material—equivalent to layering together large numbers of strips of paper as in Figure 1.6c. The starting sample is now a disc or cylinder, rather than a thin line. As the solvent moves through the column, solutes present in the original sample can do one of three things. Completely partitioned into the solvent they will run with the solvent front ($r_f = 1$) and be washed out quickly. Totally adsorbed to the matrix ($r_f = 0$) they will remain in their starting position. The usefulness of chromatography lies in the third possibility ($0 < r_f < 1$), where partial adsorption retards the solute, but eventually it is eluted. Two solutes with closely similar retardation constants ($r_f$) would be eluted with midpoints at different times; if the column is long enough and longitudinal diffusion is not too great, a complete separation can be achieved. The separation ability depends partly on the number of effective adsorptive steps in the process. Because the sample must have a finite depth, the number of processing steps, called the effective plate number, depends on the length of the column relative to the depth of the sample, as well as on the fineness of the adsorbent, plus consideration of diffusion. Since diffusion will spread any sample below a certain size to much the same amount, the width of the eluted fraction is the important feature, being dependent on both the depth of sample originally and the diffusion during passage down the column. As we shall see later (Chapter 4) column chromatography usually involves changing buffer conditions, either by a stepwise instantaneous change or by an applied gradient moving from one condition to another; in these cases the concept of plate number is less relevant.

The wide variety of adsorbents suitable for proteins has made column chromatography the most important basic procedure for enzyme separation. Some adsorbents have very high capacities for certain proteins ($>100$ mg cm$^{-3}$), whereas others may have capacities 200-fold less than this; consequently the *size* of column required may vary substantially. Theoretically, for an ideal system the *shape* should not matter, since the ratio of length to sample depth remains the same if the volumes of adsorbent and sample are the same (Figure

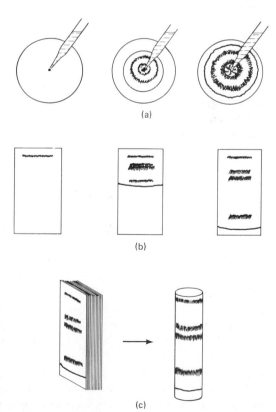

**Figure 1.6.** Principles of chromatography. In (a) the sample applied at the center of a filter paper disc spreads radially as more buffer is added, and the components separate in rings according to their $r_f$ value. In (b) one-dimensional paper chromatography results in separation of components in bands. (c) Illustration of the extra dimension acquired in column chromatography.

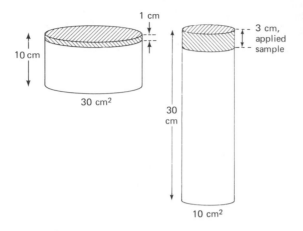

**Figure 1.7.** Two columns of identical volume but differing proportions.

1.7). But particle size is also relevant; sharper separations are obtained if more adsorbent particles are "passed" by the protein bands—giving a higher plate number—so the longer column has an advantage for a given sized particle. On the other hand a fatter column can use finer particles, since the slower flow rate per square centimeter that is usually necessary with fine, closely packed adsorbent is compensated for by the large cross section. In practice, uneven flow and non-ideal operation mean that a longer column is likely to give a better separation (see Figures 5.8 and 5.9). But it will take proportionately longer to run, which could be disadvantageous. Columns with capacities from 1 cm$^3$ to approximately 2 liters should be available in the laboratory, with the possibility of obtaining any given capacity either in a squat form where diameter $\simeq$ height, to a long form whose diameter $\simeq$ $\frac{1}{30}$ height; there are uses for all shapes and sizes.

The best columns have carefully designed surfaces for the top and bottom of the packing material so that even flow down the cylinder of material is not disturbed at either end. An efficient porous net or disc at the bottom allows liquid to flow through to the collecting space—which should be as small a volume as possible to minimize mixing. Similarly on the application end the sample and subsequent eluting buffers must be distributed evenly over the surface of the material (Figure 1.8).

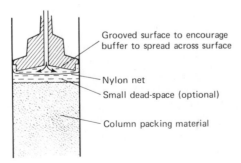

Grooved surface to encourage buffer to spread across surface

Nylon net

Small dead-space (optional)

Column packing material

**Figure 1.8.** Typical design of applicator for a chromatography column.

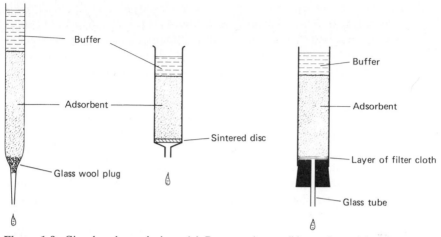

**Figure 1.9.** Simple column designs: (a) Pasteur pipette, (b) a syringe, (c) a glass tube, all adapted as columns for chromatography.

In a teaching laboratory with dozens of students it is not economically possible to give each student such a column. Simple cheap systems include (for small scale) a Pasteur pipette with a plug of glass wool in its constriction (Figure 1.9a); a syringe either with or without the plunger, with a sintered plastic disc placed at the bottom (Figure 1.9b) or a simple glass tube with a rubber stopper; and some filter material such as a sintered disc or layers of filter cloth to assist in the smooth collection of eluant (Figure 1.9c). Uneven flow onto the

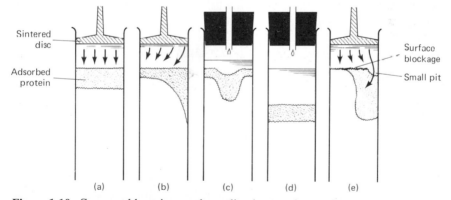

**Figure 1.10.** Some problems in sample application to columns: (a) Ideal situation; the flow is evenly distributed over the surface. (b) Due to a small gap in the side of the applicator sinter disc, most liquid flows through that gap and uneven flow onto the adsorbent occurs. (c) Drops from tube application create pit in adsorbent resulting in excessive flow in the center of the column. (d) Situation in (c) is remedied by having a large reservoir of buffer to absorb local disturbances. (e) Particulate material in sample results in clogging of adsorbent surface. If a small pit occurs in one side of the surface, all the flow may pass through it, causing highly irregular result. Remedy— centrifuge or filter sample before running on to column.

surface of the packing material can create very uneven patterns of flow through the column; a small layer of liquid above the material can help to even out irregularities, provided that the adsorbent is a material that packs firmly and is not disturbed by buffer turbulence above it. Some examples of causes of irregular flow are illustrated in Figure 1.10. As many protein solutions are colorless, it is often not appreciated how irregularly a column might be working.

A second cause of irregularity is due to uneven packing of the column. The column material is virtually always of a range of particle sizes; this means that undisturbed settling will result in the largest particles falling to the bottom, and the top part of the column would consist of the smallest. This in itself may not matter too much except for reproducibility considerations. On the other hand, disturbed flow during settling may set up irregularities which result in a column with patchy particle distribution and patchy packing, causing uneven flow characteristics. Ideally a column should be poured in one step with continuous stirring of the slurry, as illustrated in Figure 1.11. Short columns can conveniently be poured using a thick slurry, not more than 1 vol of buffer to 1 vol settled packing material, allowing it to settle rapidly with the exit tube

**Figure 1.11.** Ideal method for pouring column; the slurry is continuously stirred in the reservoir above the column to ensure that particles of all sizes are settling throughout the column.

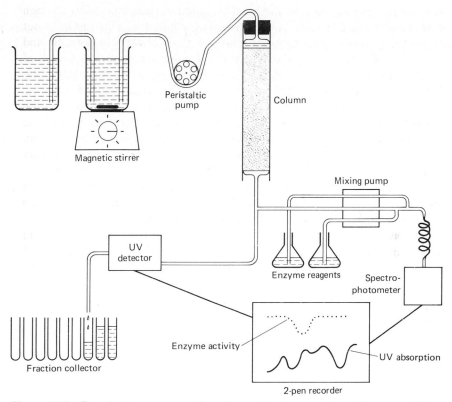

**Figure 1.12.** Complete arrangement for column chromatography, with automatic collection of eluate in fraction collector and continuous monitoring of enzyme activity.

open. After pouring the column, several volumes of buffer should be allowed
to pass through to ensure equilibration; this is particularly important with ion
exchangers, where the slurry pH should be adjusted to the correct value before
pouring.

The flow through a column can be under gravity; if a meter or so of water
pressure does not result in a sufficiently fast flow, then the system is probably
partially blocked or the tubing used is too fine and causes a resistance. Peristaltic pumps which deliver a constant rate of buffer are used widely, and
enable the sample, buffer, and gradient former to be placed at a convenient
height.

Separation of proteins is achieved in the column, and the results are monitored as soon as the liquid exits from the column. The remaining apparatus,
shown in Figure 1.12, is for convenience—continuous monitoring of UV
absorption at 280 nm for protein (Trp + Tyr band), or more sensitivity at
lower wavelengths 220 nm → 205 nm (peptide band), and finally collecting
fractions of equal volume using a timed control.

If a peristaltic pump is not used, flow rates could change, and drop counting may prove more satisfactory for getting equal fractions. On the other hand, changes in surface tension as proteins emerge can result in different-sized drops. The flow from the column may be split, with a small amount being used for a continuous enzyme assay, which if synchronized properly could record the enzyme activity superimposed on the protein curve. Gradient elution can be accomplished using a gradient former which may be simply two beakers connected by suitable tubing for siphoning. Somewhat more reliable and reproducible gradients are formed using two cylinders connected at the bottom, with an efficient paddle stirrer (see Figure 4.19). Finally, the "black box" type of gradient former can be used to generate automatically any sort of gradient required.

## 1.4  Manipulation of Protein Solutions

As a final section in this introductory chapter on general methods, some notes on the handling of protein solutions are presented. Special techniques for maintenance of enzyme activity and stability are described in Chapter 6. Very often you have a solution of protein which is not in the right state for the next fractionation procedure. It might be in the wrong buffer, at the wrong pH, have too much salt, be too dilute or too turbid. Turbidity is indicative of particulate material, which can be centrifuged or filtered as described in section 1.2. The pH can be changed readily using a pH meter and not too strong an acid or base (cf. section 6.1). Concentration of solutions and ways of changing buffers are described now in more detail.

### Concentration

Concentrated solutions can be diluted easily, but the second law of thermodynamics interferes with the converse. Concentration of protein solutions can be done in a number of ways, and the best way will depend on the particular application and requirement. Concentration can be carried out by precipitation, followed by redissolving in a small volume. Typically this would use ammonium sulfate, though acetone is sometimes useful (cf. section 3.4). The redissolved precipitate now contains precipitant, which will probably need removing. Precipitation methods are suitable only from a protein concentration of above 1 mg ml$^{-1}$; lower concentrations either fail to precipitate if the solubility is not low enough or denature during the aggregation process. However there are many exceptions to this statement, as some proteins are much more stable than others in dilute solution.

Adsorption, e.g., to an ion exchanger, may be appropriate and can achieve successful concentration of very dilute protein solutions. (As long as the required enzyme is totally adsorbed, it does not matter much whether other proteins are.) For example, a very dilute protein solution eluted from DEAE-cellulose at low ionic strength pH 6 would adsorb to a fresh, small column of

DEAE-cellulose if its pH is raised to 8, and then could be eluted with a small pulse of salt.

The most important methods for concentrating dilute protein solutions involve simply removing water (plus low-molecular-weight solutes). This can be done by a variety of procedures based on the semipermeable membrane (dialysis) or gel filtration principle, in which proteins cannot pass through a membrane or surface that water and small molecules can. The most commonly employed system is that known as ultrafiltration, in which water is forced through the membrane, leaving the more concentrated protein solution behind. This is done either by reducing the pressure to "suck" the water out (Figure 1.13a) or more commonly using compressed air or nitrogen at up to 5 atm in a pressure cell, which is constantly stirred to prevent clogging of the membrane (Figure 1.13b). The latter, when operating efficiently, can reduce 50 ml to 5 ml, or 200 ml to 10 ml within an hour or less, especially if the final concentration is still fairly low; ultrafiltration speed drops off rapidly as protein concentration rises.

Dry gel filtration particles which exclude protein when swollen can be employed for rapid concentration. Since it is not possible to remove all protein solution from the outside of the particles without washing (causing a redilution), this method is only really suitable if concentration of the solution is more important than overall recovery; at best a 10–15% loss of protein can be expected. For instance, 100 ml of solution could be treated with 20 g *dry* Sephadex G-25. The powder rapidly swells as it takes up water, eventually occupying almost the whole volume. The slurry can then be filtered under suction, obtaining about 30 ml of solution containing some 80% of the original pro-

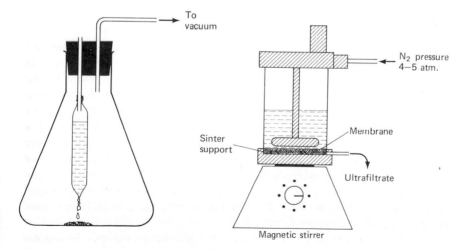

**Figure 1.13.** Ultrafiltration of protein solutions. In (a) the suspended dialysis bag is open to atmospheric pressure, while outside it is evacuated—maximum pressure difference less than 1 atm. In (b) the ultrafiltration cell can be operated at up to 5 atm positive pressure, forcing filtrate through membrane.

tein—approximately a twofold concentration. This is very useful if rapid concentration of an already concentrated protein solution is desired, but is not generally used in enzyme purification procedures because of the losses involved.

The final method of concentration by water removal involves osmotic forces. The sample is placed in dialysis tubing, which is then immersed in a solution, or surrounded by a powder, which attracts water from the bag. Commonly used substances are polyethylene glycol (MW 20,000+) and highly substituted carboxymethyl cellulose. Both of these polymers attract water very strongly, maintaining a low water activity outside the bag and high inside, forcing the water out. Calbiochem has adopted the trade name of Aquacide for these products. On a small scale it is very convenient: 10 ml can be concentrated to practically nothing in an hour by dabbling the dialysis bag in powder, occasionally stripping off the hydrated gel forming on the outside. For large amounts a concentrated solution (say 20%) of polyethylene glycol can be used, and the dialysis bag left, with stirring, for hours without attention. Provided that the polymers do not contain low-molecular-weight impurities damaging to the enzyme, the method is quite satisfactory.

## Removal of Salts; Changing Buffers

Concentration of protein solutions and dialysis are closely related. Conventional use of dialysis bags involves the removal of unwanted low-molecular-weight solute from the sample and replacement with buffer present in the "dialysate." Osmotic forces are usually the opposite to that described above: high concentrations of salt or organic solvent in the sample cause water to enter the bag before the salt leaves; consequently there is an increase in volume during the early stages if the solute concentration is high. Dialysis is used both for removing excess low-molecular-weight solute and simultaneously introducing a new buffer solution (it may be just water) to the sample. Dialysis tubing is available in a variety of sizes, and conveniently does not allow molecules larger than 15,000–20,000 daltons to pass through. All low-molecular-weight molecules diffuse through the tubing, and eventually the buffer composition on each side equalizes. Complete removal of endogenous salts from a sample cannot occur in one dialysis simply because at equilibrium what was originally inside the dialysis bag now is distributed throughout the buffer *and* the dialysis bag. If the buffer volume is 50 times that of the bag, then the best that can be achieved is a 51-fold dilution of the original salt/low-molecular-weight material. Consequently, changing the buffer at least once is needed, and to speed up the process, stirring of the buffer *and* movement of the dialysis bag should occur—a well-mixed system can reach more than 90% equilibration in 2–3 hr. A large volume of buffer is preferable since fewer buffer changes are needed (Table 1.1). Dialysis is a convenient step to be carried out overnight as it needs no attention, and equilibrium is reached by morning. However, as will be discussed later (cf. section 7.1), the possibility of proteolytic degradation during dialysis may make its use undesirable.

There are methods for pretreating dialysis tubing to remove a variety of chemicals introduced during manufacture. Boiling with an EDTA-NaHCO$_3$ solution has been recommended, and long-term storage immersed in such a solution gives a good result. Chemical contamination problems are more likely to arise when dialysing a very dilute protein solution, since then the membrane/protein ratio is high. Protein solutions of the order 10 mg ml$^{-1}$ and more are unlikely to be affected; in practice it is usually safe to use the tubing direct from the dry roll. After wetting, a knot (preferably two) is tied in one end and a filter funnel placed in the other for pouring in the sample. With the end of the tubing resting on a convenient surface, the sample is poured in (Figure 1.14). It is best to have the tubing flat to start with so that air does not have to get out through the funnel. The tubing is then closed with a further knot and placed in the dialysis buffer. Remember to leave space for expansion if there is a high solute concentration in the sample. Otherwise expansion will cause swelling and a high pressure inside the bag. This might result in bursting, or else leakage through the knots.

Before gel filtration techniques became widely used, dialysis was a routine standard procedure, involved in almost all enzyme purifications. Although it is simple and requires little equipment, it has two major disadvantages: the need to change dialysate perhaps at awkward hours and the relative slowness of the method. The importance of speed is discussed later (cf. section 7.1). Except when it would be a case of carrying out dialysis overnight compared with not doing anything overnight, the rapidity and complete separation achieved by gel filtration makes this a much preferable method—especially on a small scale.

The principles of gel filtration are well known and need not be detailed here.

**Table 1.1.** Dialysis Protocols for Decreasing Salt Concentration from 1 M to <1 mM[a]

---

*Procedure 1*
Dialysis against 1 liter of water                              Complete equilibrium
  → swells to 100 ml, at equilibrium = 95 mM          would take 3–5 hr
Change dialysate, further 1 liter of water                     at least; with
  → swells to 110 ml, at equilibrium = 9.3 mM         shorter times, a
Change dialysate, further 1 liter of water                     fourth change
  → no further swelling, at equilibrium = 0.9 mM      would be needed.

*Procedure 2*                                                  A single change
Dialysis against 5 liters of water                             would be sufficient
  → swells to 110 ml, at equilibrium = 20 mM         even without
Change dialysate, further 5 liters of water                    complete
  → no further swelling, at equilibrium = 0.4 mM      equilibrium.

---

[a]Initial sample is protein in 50 ml of buffer containing 1 M ammonium sulfate (redissolved precipitate from an ammonium sulfate fractionation).

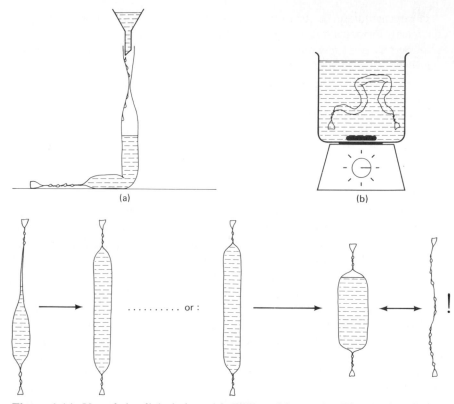

**Figure 1.14.** Use of the dialysis bag: (a) Filling with sample, (b) agitation during dialysis, (c) results of swelling due to osmotic forces.

For removing salts and buffer changing, the "all or none" principle of gel filtration is involved; the medium chosen totally excludes proteins, but includes small-size solutes. Probably the most widely used material is Sephadex G-25; Biogel P-30 has approximately the same properties. Packed into a column pre-equilibrated with the buffer required in the sample, rapid "desalting" can be achieved in one pass, provided that the sample volume is no greater than one-fifth that of the column. Satisfactory separation is achieved at flow rates of up to 50 cm hr$^{-1}$ (*note:* 50 cm hr$^{-1}$ equals 50 ml hr$^{-1}$ per cm$^2$ cross-section). A range of prepacked columns should be on hand so that changing the buffer or desalting a solution can be a routine and immediate operation. Table 1.2 lists the range that we routinely use. Small columns should be packed with finer-grade beads to retain optimum resolution.

The sample should not have too high a concentration of either salts or protein if optimum separation is needed. After dissolving an ammonium sulfate precipitate with a *minimum* amount of buffer, it should be diluted with at least a further equal volume of buffer before desalting on a column. Protein concentration should not be more than 30 mg ml$^{-1}$. Protein emerging from the column, if it is not colored and not being monitored continuously in a UV monitor, can be detected by spot tests using trichloracetic or perchloric acid precipitation. Ammonium sulfate can be detected by spot tests using barium chloride. (If the column buffer contains phosphate, use barium chloride in HCl to avoid confusion with barium phosphate precipitates.) Remember that a gel filtration column should be preequilibrated with the buffer that the protein is to be exchanged into, using at least one column volume before applying the sample; but it does not matter what the column is washed with, since the protein travels ahead of the solvent front. When using expensive buffers, rather than equilibrate the whole column with buffer, one can use a "desalting buffer" such as a very low concentration of Tris-EDTA. About 0.2 mM of the acid form of EDTA is neutralized to pH 7.5–8 with Tris; this has no effect on the proteins unless (a) EDTA inactivates by pulling out an essential metal ion, or (b) the proteins "salt in" (cf. section 3.1), at very low ionic strength and so precipitate in the column. Provided that these circumstances are not occurring, the emerging protein can be made up to the correct buffer composition by, for instance, adding 0.11 vol of a 10× concentrated stock solution.

As with dialysis, the final volume is likely to be considerably larger than the initial volume before "desalting"; nevertheless, further dilution might be required before application to an adsorption column (Chapter 4). Changing of buffers, rather than removal of excess salt, is carried out in exactly the same way; however, if the protein concentration is already low and the volume large, it may be better to "salt out" first, dissolve in a small volume of buffer, and then use a relatively small desalting column.

This chapter has described some of the basic operations in the enzyme purification laboratory. Chapters 3, 4, and 5 go into the details of fractionation procedures used for isolating enzymes. But first, one needs an extract of the raw material to work with; obtaining this extract is described in the next chapter.

**Table 1.2.** Dimensions of Columns for Desalting Using Sephadex G-25

| Column dimensions, cross-section area (cm$^2$) × height (cm) | Resin grade | Column volume (cm$^3$) | Sample size (cm$^3$) | Normal flowrate (ml hr$^{-1}$) | Typical time for completion (protein emergence) (min) |
|---|---|---|---|---|---|
| 1 × 8 | (fine) | 8 | 1–1.5 | 40 | 7 |
| 6 × 10 | (medium) | 60 | 2–10 | 200 | 15 |
| 8 × 30 | (medium) | 240 | 10–30 | 250 | 30 |
| 8 × 60 | (medium) | 480 | 30–80 | 250 | 45 |
| 16 × 90 | (coarse) | 1440 | 80–300 | 500 | 90 |
| 50 × 100 | (coarse) | 5000 | 300–1000 | 1500 | 90 |

# Chapter 2
# Making an Extract

## 2.1 The Raw Material

To many people embarking on an enzyme purification project, much of this section will not be relevant, for they have no choice of raw material. A wide variety of different circumstances could prevail; the laboratory may be working solely on a particular organ or species, and the only considerations about the raw material may be questions of availability in sufficient quantities at times when needed, and the possibility of frozen storage before use.

At the next level the researcher may have some choice of tissue to use for purifying the enzyme, or alternatively the tissue/organ may be defined but the species source allowed some variation. Species used for enzyme purification have been decided principally on the ease of raising/growing the material. With animals the principal species used are the rat (especially for liver studies), rabbit (especially for skeletal muscle) and meat animals, mainly beef and pig, for organs which are rather small in size in the laboratory animals (e.g., heart, brain, kidney, thymus). In addition, one must not forget the human animal, on behalf of whom so many laboratory animals are sacrificed, on the assumption that what is true for the rat is likely to be true for the human. Studies on enzymes from human tissue have of necessity been limited. On the other hand, human blood is readily available, and there are many procedures for isolating enzymes from erythrocytes and other blood cells; plasma contains relatively few true enzymes but many nonenzymic proteins. More recently enzyme studies on a wider variety of species have been commoner, partly through the realization that many important features of metabolic control, enzyme localization and characteristics do differ markedly even within the vertebrate phylum.

Invertebrates' enzymes have been rather neglected, mainly due to the fact that most invertebrates are very small, and a large number are required to get a reasonable amount of starting material. If the organ of interest is not dissected from each individual, an arduous task with insects, whole-body homogenates cause great problems because of the large quantity of digestive juices, including proteolytic enzymes, that are released into the extract. Most well-known invertebrate enzymes have been made from the larger crustacea such as crayfish and crabs. Many a feast of lobster claws has been enjoyed in laboratories isolating enzymes from the tail muscle!

Plant biochemistry has long been the poor relation to animal work, largely because of the medical implications of the latter, but also because plants pose particular difficulties to the enzyme chemist. The variation of quality with season can only be overcome by growing the plants in special growth chambers; otherwise one has to put up with this variation for material grown in the open. Of the vast variety of species, those studied have been, for obvious reasons, mainly those of economic importance, and these are not necessarily the best choices biochemically. Nevertheless, one or two plants have tended to predominate because of the ease of extraction of their leaves, especially when it is easy to isolate chloroplasts from these extracts. Most commonly used is spinach, *Spinacia oleracea,* or alternatively the (not closely) related silver beet *Beta vulgaris,* which is easier to grow in a range of climatic conditions. Plant cells are highly compartmented; in most cases the bulk of the volume is vacuolar space, which can be filled with quite acid solutions, proteases, and a variety of other detrimental compounds. There is also a large amount of cell wall (cellulose), and the chloroplasts, starch granules, and other organelles occupy much of the cytoplasmic space. Indeed the cytoplasm (which, together with the internal space of the chloroplasts, contains most of the enzymes) may often be no more than 1–2% of the total cell volume in plants. Consequently plant extracts may be very low in protein even if very little fluid is added when making the extract.

Microorganisms again present a different range of problems. Whereas the animal source may be available at the local abbatoir or at least in the institution's animal house, and a plant source may be at the supermarket, microorganisms (with the notable exception of yeast) will have to be specially grown in controlled conditions on a fairly large scale. Algae, fungi, yeasts, and bacteria each have special requirements for growth conditions, harvesting, and extracting. The particular problem of marked changes in enzyme composition during different growth phases must be carefully studied. Collection of bacteria and other unicellular organisms during the log phase of growth is usually desirable (Figure 2.1), though the enzyme being studied might not be at a maximum activity at this time; preliminary investigations on the organism to determine what physiological state contains the most of the enzyme required should be carried out. *Saccharomyces cerevisiae,* baker's yeast, is so readily available in large quantities that if this can be used the whole problem of raw material is solved. Compressed yeast cakes are sold which represent nearly 100% pure yeast cells at the end of their growth phase. Grown on a variety of nutrients, the main carbon source is usually molasses or another sucrose source. Yeast cakes remain viable (and so the enzymes remain active) for some weeks at $0°C$, and if frozen will be usable months later. Also very popular with bakers and biochemists are the cans of dried yeast which reconstitute on adding water. These are convenient sources for many enzymes, though the drying process may tend to partly inactivate a few of the more sensitive ones. Sealed under vacuum or in nitrogen and stored at 0–4°C, these cans can be used years later, and represent perhaps the most convenient raw material source there is—provided that you can break open the cells (see below).

The true enzymologist is one who is rarely concerned with where the enzyme comes from, but only wants as much as possible of it from the most convenient source. In this case he should study the literature and determine the best organism which he can get hold of conveniently; if comparing results from different laboratories, it may be worth doing some exploratory measurements himself. The most important thing is to be sure that all the enzyme present in the original material can be extracted, so that the amount per weight of tissue can be determined reliably. If the cells are not broken properly, an otherwise excellent source might be overlooked because much of the activity was not released but removed with the residue while making the extract. And if bacteria are to be the source of the enzyme, the experimenter should consider the possibility of selecting out strains that have abnormally high levels, either by conventional mutant selection or by more sophisticated cloning techniques whereby multiple copies of the enzyme's gene are inserted via plasmids into the bacteria's genetic complement (1). The expression of the gene can be further enhanced by fusing to a strong promoter (2). This may take a considerable time to achieve and would need specialist expertise and handling; it is not worth attempting if the time allotted for isolating the enzyme is short.

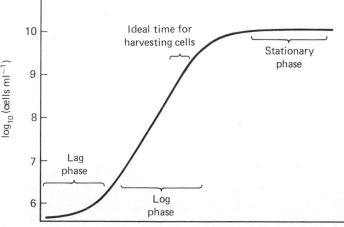

**Figure 2.1.** Growth of microorganisms in nutrient-rich medium. The ideal time for harvesting cells is toward the end of the log phase before growth rate slows, giving a high yield of cells.

## Freshness and Storage

Usually the sooner the raw material is used, the better and more physiologically relevant the preparation will be, for if the tissue has been "dead" for some time, natural degradative processes will have commenced. Yet there are occasions when absolute freshness can be a nuisance. For example, skeletal muscle contains a high level of ATP, with phosphocreatine and glycogen present which maintain that level for some hours after death. The structural proteins myosin and actin are normally insoluble at ionic strengths below about 0.25, but the presence of ATP causes the disrupted myofibrils to swell, and a portion of these proteins go into solution. Even at physiological ionic strength ($\sim$0.16) some myosin can be solubilized. This can subsequently upset enzyme preparations by precipitating and mixing into fractions which should not contain it. A particular example is the preparation of AMP deaminase (3). Extraction with a buffer of ionic strength 0.25 is recommended—with really fresh muscle much actin and myosin go into quasi-solution and upset the next step, adsorption on phosphocellulose. But if the muscle is allowed to go into rigor mortis first, indicating a loss of ATP, a much better preparation is obtained. This is a rare example; freshness should always be the prime aim unless there is some good reason for delaying things.

The availability of fresh material does not always coincide with one's ability to use it. Frozen storage, either of the material as received or of an extract from it, must be considered. During freezing a great many events occur. Firstly the free water freezes and ice crystals grow. The ice crystals are very destructive for membranous layers and organelles, but do not normally upset enzymes directly. As the temperature approaches the eutectic points of the various salts present, salts crystallize out also. The less soluble of a pair of salts comprising a buffer will come out of solution first; consequently the pH can change drastically before complete solidification takes place (4,5). If the storage is at a temperature of between $-15°C$ and $-25°C$ (typical of the temperatures in domestic refrigerator freezing compartments and simple freezers), the remaining concentrated solution, perhaps at a quite different pH from the original solution, remains unfrozen. Proteases liberated from lysosomes during the freezing can go to work (albeit slowly at such low temperature) in the concentrated protein soup, and a few weeks of storage may do a lot of harm. Thus freezing should aim at reaching a temperature below $-25°C$ quickly, and storage at even lower temperature if possible. Commercial freezers operating at temperatures down to $-80°C$ are available.

Freezing of extracts is sometimes preferable, because the composition of the medium can be manipulated to optimize storage conditions. Proteolytic enzyme inhibitors (cf. section 6.2) can be added after making the extract, and the pH adjusted with a suitable buffer to the best value for stability of the enzymes. Also, large amounts of raw material can all be processed into one extract, so that subsequent samples taken from frozen storage are identical except for their storage time; variation in behavior cannot then be blamed on biological variation of the material frozen away.

Remember that thawing speed can be important—the faster, the better, provided that local overheating does not occur. The best way is to immerse the container in warm (40–50°C) water and agitate frequently. It is unlikely that in these conditions the melted solution would rise above 10°C, despite the outside temperature, as long as ice remains inside.

Now that we have the raw material, either fresh or thawed, it is time to homogenize it and obtain an extract containing the enzyme in solution, the first step in the majority of enzyme purification procedures. Cases where preparation of subcellular particles or other insoluble material precedes the release of the enzymes into solutions are described briefly later (cf. section 2.4).

## 2.2   Cell Disintegration and Extraction

Most of the proteins and enzymes studied in the early days of protein chemistry were isolated from extracellular fluids. The reason for this is not just because it was easy to obtain the raw material, but because extracellular proteins are for the most part more stable, often as a result of disulfide cross-links, and they tend to be small molecules; early studies on protein structure naturally concentrated on such small proteins. Thus lysozyme, ribonuclease, and chymotrypsin were among the earliest proteins studied in detail, and all are from extracellular sources. But most enzymes are found inside cells, and are very often much less stable; disulfides are generally absent because of the more reducing intracellular environment. The purpose of the present section is to describe methods of disrupting the cells and releasing the enzyme into an aqueous "extract" which is the first stage of enzyme purification techniques.

There are many methods of cellular disintegration, for there are many types of cell. Most cells have particular characteristics which need special attention during disintegration. Animal tissues vary from the very easily broken erythrocytes to tough collagenous material such as found in blood vessels and other smooth-muscle-containing tissue. Plant cells are generally more difficult to disrupt than animal cells because of the cellulosic cell walls. Bacteria vary from fairly fragile organisms which can be broken up by digestive enzymes or osmotic shock to more resilient species with thick cell walls, needing vigorous mechanical treatment for disintegration. It is generally not advisable to use a disruption treatment more vigorous than necessary, since labile enzymes may be inactivated once liberated into solution. Table 2.1 gives a list of techniques that can be used, with illustrations in Figure 2.2.

These methods are for disrupting cells and releasing enzymes into solution. On occasions when the enzyme is present in an organelle, the methods may still be suitable in that they also disrupt the organelles. On the other hand, it may be desirable to isolate the organelles themselves first, so disposing of contaminating cytoplasmic proteins, before extracting the enzyme from the organelle. Less vigorous techniques are needed; preliminary digestion of collagenous or cellulosic extracellular structures with proteolytic enzyme preparations enables the cells to be broken subsequently by a gentler treatment while maintaining the integrity of the organelles. Preparation of mitochondria from tissues such as skeletal or cardiac muscle using proteinase treatment is an example of such

**Table 2.1.** Cell Disintegration Techniques

| Technique | Example | Principle |
|---|---|---|
| *Gentle* | | |
| Cell lysis | Erythrocytes | Osmotic disruption of cell membrane |
| Enzyme digestion | Lysozyme treatment of bacteria | Cell wall digested, leading to osmotic disruption of cell membrane |
| Chemical solubilization/ autolysis | Toluene extraction of yeast | Cell wall (membrane) partially solubilized chemically; lytic enzymes released complete the process |
| Hand homogenizer | Liver tissue | Cells forced through narrow gap, rips off cell membrane |
| Mincing (grinding) | Muscle etc. | Cells disrupted during mincing process by shear force |
| *Moderate* | | |
| Blade homogenizer (Waring-type) | Muscle tissue, most animal tissues, plant tissues | Chopping action breaks up large cells, shears apart smaller ones |
| Grinding with abrasive (e.g., sand, alumina) | Plant tissues, bacteria | Microroughness rips off cell walls |
| *Vigorous* | | |
| French press | Bacteria, plant cells | Cells forced through small orifice at very high pressure; shear forces disrupt cells |
| Ultrasonication | Cell suspensions | Micro-scale high-pressure sound waves cause disruption by shear forces and cavitation |
| Bead mill | Cell suspensions | Rapid vibration with glass beads rips cell walls off |
| Manton–Gaulin homogenizer | Cell suspensions | As for French press above, but on a larger scale |

a procedure (6). However, this is generally only applicable on a small scale, more suited for metabolic studies on the organelles than enzyme purification. Yields of purified organelles may be very low, and it is often a better procedure to do a complete tissue disruption and then approach the problem of isolating the enzyme required from the complex mixture of proteins in the extract. Thus mitochondrial and chloroplast enzymes would often be prepared from a complete tissue homogenate rather than from the isolated organelles.

Finally, the enzyme may not be soluble in the extraction buffer. In this case special techniques are required and are discussed briefly in section 2.4.

The "extract" is prepared, after cell disintegration, by centrifuging off insoluble material. Before centrifuging, the mixture is usually described as an homogenate; after centrifugation as much as possible of the enzyme should be present in the liquid layer. Liquid is trapped within the precipitated residue, and the total loss will be related to the proportion of residue to liquid. When making an extract, losses due to trapping within the residue should not be more than about 20%, unless the raw material is very readily available, and high concentration/low volume of extract is more important than recovery from the homogenate. Since you can expect about half of the volume of the residue to be trapped liquid, simple calculation indicates that the volume of supernatant should be about twice the volume of residue to get 80% recovery. Most animal tissues have large amounts of insoluble cell material which bind a lot of water. They give a volume of residue about as much as the original tissue (dependent somewhat on the amount of centrifuging). Thus when making a homogenate of, say, liver, *at least* 2 vol of a suitable extractant buffer should be added when making the homogenate. The more that is added, the larger proportion of the soluble fraction will be extracted, but the extract will be more dilute; its greater volume may create difficulties when working on a large scale. Two-and-one-half vol of extractant is typical for liver, heart, or skeletal muscle homogenates.

Plant tissues are quite different; as mentioned in the previous section, only a small fraction of the volume of plant tissue is truly intracellular; large vacuoles (being regarded here as extracellular) and intercellular spaces mean that on disruption much liquid is released, making additional extractant liquid almost unnecessary. The residue after centrifugation may occupy only 20–40% of the volume of the original plant tissue. Nevertheless, it may be important to use some added extractant liquid so as to control undesirable processes during homogenization. These include acidification and oxidation of susceptible compounds. A particular problem with many plants is their content of phenolic compounds, which oxidize—mainly under the influence of endogenous phenol oxidases—to form dark pigments. These pigments attach themselves to proteins and react covalently to inactivate many enzymes. Two approaches to this problem are useful. Firstly, the inclusion of a thiol compound such as $\beta$-mercaptoethanol minimizes the action of phenol oxidases. Secondly, addition of powdered polyvinylpyrrolidone is often beneficial as it adsorbs the phenolic compounds.

Microorganisms are more like animal tissue with respect to the volume of residue after centrifuging; yeasts and similar fungi with thick cell walls give a residue which nearly equals the volume of cells initially. Bacteria may do so also; however, if the method of extraction involves vigorous shear forces, much of the cell wall material may actually go into solution, creating further problems which are discussed below (cf. section 2.3).

Cell debris may centrifuge down easily, particularly the coarse residues from animal and plant tissues, although fine intracellular particulates will remain in suspension after, say, $10,000 \times g$ for 15 min. This cloudiness need not be of any concern, as fractionation steps will remove it. But use of the more vigorous disruption techniques such as bead-milling can break down cell wall material into fine particles which may have a relatively low density, since lipids are usually present. This can be difficult to sediment at the available $g$ force

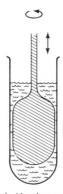

(a) Hand-operated
or motor-driven

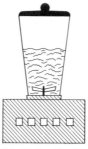

(b) Waring blender

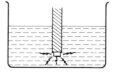

(c) Ultrasound

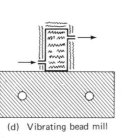

(d) Vibrating bead mill

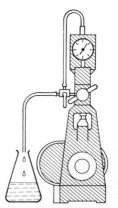

(e) Manton-Gaulin homogenizer

**Figure 2.2.** Equipment used for breaking up cells to obtain an extract: (a) hand-operated or motor-driven glass homogenizer, (b) Waring blade-blender (food processor), (c) ultrasonic probe, (d) vibrating glass bead mill, (e) Manton–Gaulin cell disintegrator.

for large volumes, and the extract may be very turbid indeed. Methods for clarification of the extract are described below (cf. section 2.3).

The following outlines some procedures for extracting various tissues.

## Mammalian Tissues

Dice tissues and cut away connective tissue and fat as far as possible. Place in Waring blender with 2–3 vol cold extraction buffer per g tissue. Blend for 30 sec, then repeat if still lumpy. Stir the homogenate for 10–15 min, checking that its pH is suitable, then transfer to centrifuge buckets. Centrifuge at 5000– 10,000 $g$ for up to 60 min (2–3 $\times$ $10^5$ $g$ min). Decant extract through Miracloth (Chicopee Mills, Inc., N.J.), cheesecloth, or glass wool to trap fat particles.

## Erythrocytes

Red blood cells are easy to extract, after collection by centrifugation, and a rinsing step in isotonic NaCl (0.9%, 0.15 M) followed by centrifugation again. The cells are osmotically lysed with water, e.g., 2 vol water to 1 vol packed cells. However, approximately 90% of the protein going into solution is hemoglobin, and unless this is what you are purifying, a method for selectively removing it is most useful. Ethanol-chloroform has been used to denature hemoglobin successfully (7).

## Soft Plant Tissues

Using only 0.5–1 vol of cold extraction buffer containing 20–30 mM mercaptoethanol, homogenize in Waring blender for 30 sec. Alternatively pass material through domestic juicer, washing through with the buffer. Centrifuge as soon as possible to minimize oxidative browning (2–3 $\times$ $10^5$ $g$ min). Decant carefully from soft material on surface of precipitate. Addition of powdered polyvinylpyrrolidone can be beneficial in adsorbing phenols.

## Yeasts

(1) Complete disruption can be obtained with a Manton–Gaulin homogenizer (Gaulin Corp., Mass.) or a Vibrogen Cell Mill (Bühler, Tübingen), using approximately 2 vol buffer per g wet weight.

(2) *Toluene autolysate.* A variety of toluene methods have been used (8–10); certain of them are not fully successful with all qualities of commercial yeast cake. The principle is to treat the yeasts with toluene, usually at a temperature of 35–40°C, when after 20–30 min the yeast "liquifies" due to extraction of cell wall components. Buffer is then added and the slurry stirred in the warm for a few hours or left overnight in the cold. As the method is autolytic, in which cell wall structure is degraded by enzymic

action, some cellular enzymes become degraded during the treatment. Ethyl acetate has been used in lieu of toluene (11,12) but does not work with tougher yeast strains.

(3) *Ammonia cytolysis* (13). This method is best suited with dried yeast; it is simple, but some enzymes unstable at pH 10 are lost completely. "Active dried" yeast is stirred with 0.5 M NH$_4$OH, 2 vol per g dry weight, at room temperature, and stirring is continued for 16–20 hr. A little toluene, 5–10% of the total volume, sometimes improves the overall yield of soluble proteins. A further 1–2 vol of water plus enough acetic acid to bring the pH down to an appropriate value is stirred in before centrifuging off the cell debris.

## Bacteria

Vigorous treatments with sonication, bead-milling or the French press successfully disrupt bacteria, though are not all convenient for large-scale extraction. Grinding with alumina is often successful on a small to medium scale. Gram-positive species are mostly susceptible to lysozyme; stirring a suspension of cells in buffer plus 0.2 mg ml$^{-1}$ egg white lysozyme at 37°C for 15 min may be sufficient to release the cytoplasmic components. Inclusion of deoxyribonuclease I (10 $\mu$g ml$^{-1}$) improves the quality of the extract obtained by reducing its viscosity.

Gram-negative species are less susceptible to lysozyme without prior treatment. A combined nonionic detergent–osmotic shock–lysozyme treatment has recently been described (14). A slight modification of this has been routinely used for extracting *Zymomonas mobilis* and *Photobacterium phosphoreum*, both gram-negative organisms; the extraction is complete, yet avoids complete dispersal of cell wall material:

To 10 g wet weight of cells, 0.1 ml of 10% v/v Triton X-100, 10 $\mu$l of mercaptoethanol, and 2.5 ml of glycerol are added. The cells are dispersed and stirred vigorously for 30 min. Then 30 ml of extraction buffer (e.g., 20 mM KP$_i$ pH 7.0 + 1 mM EDTA) containing 0.2 mg ml$^{-1}$ lysozyme and 10 $\mu$g ml$^{-1}$ DNase is added rapidly and stirring continued for 30 min. Five mg of phenylmethylsulfonyl fluoride dissolved in 0.5 ml acetone and 0.1 mg pepstatin A are added, and the suspension centrifuged (15,000 $\times$ g for 20 min).

These are just a few examples of procedures that can be used for obtaining extracts from various sources. In practice there will often be a reported method to follow, but remember that raw materials can be very variable, as can the equipment used, and one may need to modify and develop the method for particular circumstances.

Thus the basic methods of obtaining an extract are simply disruption of the raw material in the presence of a suitable buffer, followed by centrifugation to remove insoluble residues. The next section describes how the quality of the extract as a starting material for enzyme purification may be optimized.

## 2.3   Optimization and Clarification of the Extract

At this stage we will assume that a suitable method for cell disruption has been found and an extract made. It is important for reproducibility that each time an extract is made, it is as close to identical as possible. Often the properties of proteins in fractionation procedures depend not only on their own particular characteristics, but also on the composition of the solution, including other proteins present. Things that can go wrong include suboptimum operation of the homogenization apparatus—low pressure in a pressure cell, worn beads in a bead mill, or blunt blades on a Waring blender, resulting in less than complete cell disintegration. As a result the extract would be more dilute and may have a different proportional composition since some components may be selectively extracted by suboptimum disruption. It is important to know exactly how much enzyme is present in a gram of raw material to evaluate a particular extract. Probably the best method is to make a small-scale extract using a large volume (say 10 ml g$^{-1}$) of extractant, and measuring the amount of activity after different times of treatment. Extended times may eventually result in less activity because the treatment is denaturing proteins, either by heating or by the vigorous shearing nature of the disintegration method. It is very often necessary to take measures to cool the system during cell disruption; ice-cold buffer and short bursts of treatment followed by a cooling period may be needed. As mentioned in the preceding section, the volume of extractant per g material is an important consideration; on a large scale the amount used will tend to be a compromise between maximum extraction and minimum volume of extract.

Up to this stage no mention has been made of the nature of the extractant buffer, other than to suggest inclusion of things like $\beta$-mercaptoethanol for specific purposes. Plain water can be used, but it will not extract all proteins— this may be a good thing if all the enzyme one is interested in is extracted by water. Cells contain many salts and much insoluble material which is charged, e.g. proteins, phospholipids, nucleic acids. The ionic strength inside the cytoplasm of a typical cell is in the range 0.15–0.2 M; under these conditions the cytoplasmic proteins are "soluble," in that they can move around the cell. If homogenized with 2–3 vol water, the ionic strength of the homogenate may be 0.05 or less. Under these conditions the charged particulates can act as ion exchangers and adsorb proteins, especially basic ones.

There have been many reports in the literature of enzymes being distributed between soluble and particulate fractions. When homogenates were made with dilute buffers, the more buffer used, the more protein appeared in the particulate fraction (15). The ion exchange adsorbent properties (as opposed to physiologically significant, specific interactions) of cellular constituents have not always been appreciated. Thus to ensure that all "soluble" cell constituents are extracted, one should use a buffer of ionic strength similar to the physiological one, and of course at the appropriate pH. If this leads to too much extraction of unwanted compounds, a compromise may be better. Typical buffers would be 20–50 mM phosphate pH 7–7.5, 0.1 M Tris-Cl pH 7.5, 0.1 M KCl with a little buffer in it, and for isolating organelles, iso-osmotic buffers containing sucrose, mannitol, or sorbitol, as well as salts and buffer ions. Also included may be EDTA (1–5 mM), $\beta$-mercaptoethanol or cysteine (5–20 mM), and specific stabilizing agents for particular proteins, e.g., $Zn^{2+}$ for zinc-containing protein, or pyridoxal phosphate for enzymes using this as cofactor.

After the cells have been disrupted in the appropriate buffer, it may be desirable to check the pH of the homogenate. Although really fresh material, homogenized in the appropriate buffer, will give a homogenate of consistent pH, that value may well change downward due to metabolic processes leading to acidification. The best example is that of skeletal muscle, in which glycogen can be rapidly converted to lactic acid, and the pH of a homogenate could drop from 7.0 to 6.0 in 30–60 min (including during centrifugation). In such circumstances the pH can be raised *above* the desired value (e.g., with M Tris) before centrifuging. Alternatively, inclusion of a glycolytic inhibitor can arrest the process; fluoride (10–30 mM) is sometimes used.

These general comments refer to maximizing extraction conditions of all soluble components. For a particular enzyme it may be better to use other conditions, purposely preventing a complete solubilization of all components provided that the enzyme required is totally extracted.

Often the extract obtained is turbid, perhaps with fat particles floating on the surface after centrifugation. The fat can be removed by coarse filtration through a plug of glass wool or a fine mesh cloth. The particulate material in suspension will include organelles and membrane fragments which may only sediment fully at about $100,000 \times g$, an impractical procedure for liters of extract. Filtration will usually be useless, since if the filter is fine enough to trap the particles it will rapidly clog. However, if the cloudiness is relatively slight and a clear extract is needed—for instance, for passing straight on to an adsorption column—filtration can be carried out with a filter aid such as Celite (cf. section 1.2).

Let us consider the different types of extract. From some animal tissues the particulates will not constitute a large proportion of the extract and can often be ignored; they will aggregate in, say, the first cut of an ammonium sulfate fractionation and be discarded then. On the other hand, other animal tissues contain many more fats and membranous structures which end up in suspension in the extract. Acidification to a pH between 6.0 and 5.0 will usually cause aggregation so that the material can be centrifuged off at relatively low speed. Ribosomal and other nucleoprotein material is usually removed by this acidification, which can be regarded as a form of isoelectric precipitation (cf. section 3.2); the phosphate groups protonate to neutralize, or at least lessen, the charge on the particulate suspension. Provided that the enzyme wanted (a) does not also isoelectrically precipitate at the pH used, (b) does not adsorb to the precipitate forming, and (c) remains stable at the subphysiological pH, this treatment is most beneficial. The extract should be kept cold, and the pH lowered with a suitable acid (e.g. M-acetic—cf. section 6.1). After stirring for 10–20 min, the precipitate is centrifuged off and the supernatant pH readjusted if required before commencing the first fractionation step proper. Plant tissues are more acidic in the first place, and particulate material such as chloroplast fragments will not aggregate so readily. Otherwise there is a relatively small amount of particulate material in plant extract, apart from coarse particles like starch, which sediment down easily after making the extract.

Microorganisms cause a variety of problems with their extracts. Firstly, there is the large amount of nucleic acid material extracted from rapidly proliferating cells. Secondly, during cell disruption the cell wall material may either be finely dispersed to give a turbid extract or partially solubilized so that as well as protein and nucleic acids there is a large amount of gumlike polysaccharide in solution. This can cause considerable problems in the early steps of fractionation.

Bacterial extracts made with the French press, by ultrasonication, or with lysozyme are viscous due to DNA and contain much ribosomal material. All nucleic acids can be precipitated by forming an aggregate with a polycationic macromolecule; protamine from salmon and clupeine from herring milt are convenient reagents for this, since their natural function is to bind with DNA. Protamine sulfate is the usual material to use—beware, it is strongly acidic and should be dissolved in water and neutralized before use. The total amount needed should be tested; it may be as much as 5 mg per g raw material. DNA, ribosomal nucleic acid, tRNA, mRNA, and other forms of nucleic acid are precipitated, and some proteins may also be adsorbed to the precipitate. Indeed, protamine precipitate adsorption has been used as a step in enzyme purification (16,17). If it proves unnecessary or undesirable to use protamine, DNA viscosity can be reduced by inclusion of DNAase when making the extract. Ribosomes can be precipitated using streptomycin; although the amount of nucleic acid material removed by streptomycin is much less than if using protamine, little protein is lost by adsorption to streptomycin precipitates.

With nucleic acids removed, carbohydrate capsular gum may remain. This can be a problem, interfering with normal protein precipitation methods. In the presence of such material, use of ammonium sulfate or other precipitation methods as fractionation procedures may be very unsatisfactory; it can be best to totally precipitate the protein in the hope of leaving the gum in solution. Thus ammonium sulfate to 80% saturation (3.3) or acetone to about 55% (3.4) should precipitate virtually all protein. In the case of ammonium sulfate a high-speed centrifugation may be needed to pellet the protein. Even so, the redissolved precipitate still contains material that can cause irreproducible behavior on columns or other fractionation methods.

Two ways of overcoming these problems have been found recently. One is to use agarose gel "amphiphilic" adsorption (cf. section 4.8), in which all the protein is adsorbed to agarose beads at 70–80% saturated ammonium sulfate, whereas nonprotein material can be washed out. Extraction of the beads with buffer releases the protein in a solution amenable to conventional fractionation steps. The second procedure concerns the method of extraction. Lysozyme treatment does not usually release so much capsular gum—it digests it—but by itself lysozyme does not disrupt all bacterial cells. Use of lysozyme together with a nonionic detergent (14) has proved very successful in getting a relatively clean extract of bacteria, which after nucleic acid removal can be used directly for fractionation attempts. Not all bacteria cause these problems; many purification procedures are published for bacterial enzymes in which, after protamine or streptomycin treatment, conventional fractionations are successful.

Yeast extraction presents several problems, the main one being that the cell walls of fungi are mostly very thick and difficult to break open. Consequently the more vigorous methods are needed to obtain the extract; if this is to be achieved in a short time, use of a glass bead vibrating mill or a Manton–Gaulin homogenizer is the only reliable method that can be used on a large scale. These methods cause a substantial amount of low-density particulate material to be liberated, and centrifugation does not remove it all. The turbidity can upset the efficiency of precipitation by ammonium sulfate. However, it can be removed by aggregation in organic solvent (acetone, 20–25% v/v at 0°C), leaving a clear solution containing most enzymes, which can be precipitated by increasing levels of acetone (cf. section 3.4). Toluene autolysis of yeasts generally gives clearer extracts, but the conditions needed can result in proteolysis.

Strong ammonia has been used as a cytolysis method for preparing the very stable enzyme phosphoglucomutase (13). This procedure has been adapted and used extensively on "active dried" yeast, to give a clear extract containing most, but not all, glycolytic enzymes (18); some are totally lost because of the high pH. Proteolysis does not seem to be a big problem with this method, as the proteases have little activity at pH 9.5–10, the pH of the mixture. The efficiency of extraction has tended to vary from one batch of yeast to another.

At this stage one should now be ready to go ahead with the purification method; an extract containing as much of the enzyme as possible has been made, and, where appropriate, particulate and most nonprotein materials have been removed. But before concluding this chapter, there is one important class of enzymes which the preceding sections have not dealt with, that is, the insoluble, particulate-associated enzymes which by definition would not be in the clarified extract.

## 2.4  Procedures for Particulate-Associated Enzymes

This section deals briefly with methods of approaching the problem of enzymes that are structurally associated with insoluble parts of the cell, such as mitochondria, chloroplasts, plasmalemma, endoplasmic reticulum, and nuclear membranes. It does not refer to the water-soluble enzymes entrapped within an organelle, such as the mitochondrial matrix proteins, but to those embedded in, covalently part of, or strongly associated with the particulate material itself.

A substantial degree of purification over the original raw material is obviously obtained by repeated washings of the homogenate with buffers that do not solubilize the enzyme. After this, there are two approaches. One could subfractionate the precipitate by such techniques as differential centrifugation (sedimentation or density gradient), electrophoresis, or possibly "molecular sieving" of the fragments. Alternatively, the enzyme can be solubilized from the particulate material (subfractionation can, of course, precede solubilization). Not all enzymes can survive being solubilized away from their normal cellular milieu, and for these the best purification depends on how well other particulate material can be separated away from the fragments containing the enzyme.

Solubilization of a particulate-associated enzyme can be achieved by a number of different techniques. In many cases, once solubilized, the enzymes behave much like any other water-soluble enzymes and can be fractionated by the same procedures, to be described in Chapters 3–5. On the other hand, it may be necessary to solubilize the enzymes using detergents, in which case the detergent, bound to the protein, takes the place of the lipid-containing membrane. If the detergent is subsequently removed, the protein may denature, or at least aggregate and precipitate from solution.

Most membrane-associated enzymes that have been studied in detail fall in the former category; they can be released by physical, chemical, or enzymic treatments and the resulting solution fractionated conventionally. Mitochondria have provided the largest number of examples, including the many enzymic activities associated with the electron transport complex. The procedures involve (19):

(a) Disruption of the mitochondria to release matrix protein, followed by release of membrane proteins from the mitochondrial fragment by
  (i) Ultrasonication, especially at high pH, elevated temperature
  (ii) Alkali (pH 8–11) $\pm$ metal chelators (EDTA)
  (iii) Detergent solubilization
  (iv) Organic solvents–acetone powder preparation; extraction with n-butanol, ethanol, etc.
  (v) Use of phospholipase A to digest the membrane lipid.
    or
(b) A classic method, still used for some purposes, is to extract much of the lipid by making an "acetone powder" (20). The particulate suspension is treated with many volumes of acetone at very low temperature (e.g., $-20°C$ or less), leaving a dehydrated powder which can be stored for long periods. On dispersing in aqueous buffer, the proteins are solubilized from the powder.

A typical example of isolation of a membrane-associated enzyme without detergents is the preparation of the ATPase complex from beef heart mitochondria (21). The mitochondria are first isolated by a technique ensuring high yield. They are then ultrasonicated to release the matrix protein. The pH is adjusted to 9.2, with EDTA present, and sonication again carried out, where-

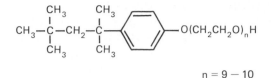

**Figure 2.3.** Structure of Triton X-100.

$$n = 9 - 10$$

upon much of the mitochondrial membrane protein goes into solution. This solution can then be treated by isoelectric precipitation (cf. section 3.2) and DEAE-Sephadex chromatography (cf. section 4.3) to yield a highly active, water-soluble ATPase.

Most membrane-bound proteins can be extracted from their membrane by detergents possessing lipophilic chains which bind to the protein at its hydrophobic surfaces in lieu of the normal membrane. Of the detergents that can be used for this purpose, sodium deoxycholate and Triton are widely used. Triton is the trade name for a series of polyethyleneglycol-based, mostly nonionic detergents, of which the most widely used for a variety of purposes is Triton X-100 (Figure 2.3), although other types have specific uses. A particularly useful zwitterionic derivative of cholic acid has recently been described (22). Detergents are capable of displacing a protein which is tightly bound by hydrophobic forces in a membrane, firstly by dissolving the membrane, and secondly by replacing the membrane by aliphatic or aromatic chains which form the lipophilic part of the detergent (Figure 2.4).

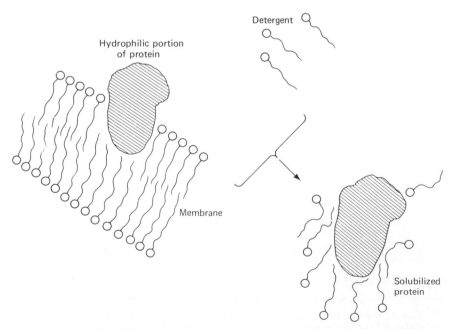

**Figure 2.4.** Action of detergent in solubilizing membrane-located proteins. The solubilized protein may be as illustrated, or bound in a more complex detergent-micelle, depending on the relative proportions of proteins and detergents.

Once the protein has been solubilized and its integrity confirmed, fractionation procedures similar to those to be described in the following chapters can be employed, and in many cases there is no particular problem. The enzyme activity must be maintained; this can be a matter of choosing a suitable detergent. Nonionic detergents such as Triton X-100 are mild in their action, and most proteins, whether originally membrane-bound or not, can tolerate Triton at levels of 1–3% w/v. At the other extreme, some anionic detergents (e.g., dodecyl sulfate) are extremely denaturing, so although efficient solubilizers, will probably not be much use in enzyme isolation.

Lipoproteins contain tightly or covalently bound lipid, which in turn is embedded in the membrane, anchoring the protein to the particulate material. Extraction with detergent may result in removal of the natural lipid, replacing it by detergent. This could cause loss of enzymic activity, restorable on addition of the natural lipid—normally a phosphatide. Such proteins are particularly difficult to deal with as they are sensitive to the detergent/lipid composition of the solution. Removal of detergent invariably results in aggregation and usually denaturation.

Excess detergent can interfere with fractionation. For instance, salting out with ammonium sulfate causes Triton X-100 to separate as a floating layer, frequently containing the proteins required, without resulting in any useful separation. Column chromatography can be carried out; gel filtration (cf. section 5.1) may separate the proteins, but detergent micelles can travel in the same area and remain with the protein. Ion exchange chromatography (cf. sections 4.2 and 4.3) is quite successful in the presence of nonionic detergents. Indeed, with normal water-soluble proteins the presence of up to 1% Triton X-100 seems to make little difference to ion exchange behavior. But membrane-solubilized proteins may be present only in detergent micelles, which influence ion exchange behavior substantially. If the protein required can be persuaded to adsorb to an ion exchanger, excess detergent will pass through; then elution of the (relatively) detergent-free protein can occur. On the other hand, if loss of all detergent results in denaturation, a small concentration (<0.1%) can be included in the buffers to prevent this; the eluted fraction will of course again contain some detergent. Nevertheless, since the object of the exercise is generally to separate a particular enzyme from other proteins, the presence of small levels of clean detergent, uncontaminated by fats, in the final preparation need be no disadvantage. Methods for removing excess of detergents have recently been summarized (23).

Further discussions of the extraction of water-soluble enzymes from membranes have been presented elsewhere (19) in comprehensive detail. Membrane-bound proteins obviously are likely to have different solubility properties compared with the typical cytoplasmic enzyme, and some unusual, empirically determined methods are occasionally employed in their purification with considerable success.

# Chapter 3
# Separation by Precipitation

## 3.1  General Observations

In the early days of protein chemistry the only practical way of separating different types of proteins was by causing part of a mixture to precipitate through alteration of some property of the solvent. Precipitates could be filtered off and redissolved in the original solvent. These procedures remain a vitally important method of protein purification, except that filtration has mostly been replaced by centrifugation. Protein precipitates are aggregates of protein molecules large enough to be visible and to be centrifuged at reasonably low $g$ forces. In some cases the aggregation continues and the precipitate flocculates, but usually the motions and collisions of the particles in suspension keep their size small. This results in clogging of filter papers and necessitates centrifugation at considerably more than 1 $g$. The various methods of obtaining a precipitate are described in the separate sections below.

Original classifications of proteins depended on their solubility behavior, and although this classification has largely been forgotten today, the words are still used daily when talking of serum albumin or immunoglobulins. Globulins were defined as proteins insoluble in low-salt solution; serum globulins can be precipitated by simply diluting serum with water, while albumins remain in solution. This globulin behavior is an example of isoelectric precipitation in the salting-in range (cf. section 3.2). The classification becomes confused since some proteins are insoluble at pH 6, but soluble in low salt at pH 8. Nevertheless there are proteins which properly fit the globulin description. They are proteins of generally low solubility in aqueous media that tend to have a substantial hydrophobic amino acid content at their surface. Conversely there are many true albumins, with high aqueous solubility and low hydrophobicity.

The distribution of hydrophilic and hydrophobic residues at the surface of the protein molecule is the feature that determines solubility in various solvents. Although hydrophobic groups tend to be concentrated around the interior of protein molecules, substantial numbers reside on the surface, in contact with the solvent (Figure 3.1). Hydrophobic groups on the surface of protein molecules are as important in determining the behavior of the molecules as are charged and other polar groups. The solvent is always aqueous-based, though there is reason to suppose it would be possible to isolate integral membrane proteins using nonaqueous solvents. The solvent properties of water can be manipulated to alter protein solubility by changes in ionic strength, pH, addition of miscible organic solvents, other inert solutes or polymers, or combinations of these together with temperature variation. The most widely used procedure is the use of neutral salts to cause selective precipitation. This is the subject of the next two sections.

## 3.2  The Solubility of Proteins at Low Salt Concentrations

Most enzymes exist in cell fluids as soluble proteins, often in a concentrated soup that may be as much as 40% protein (24). Thus they are very soluble in physiological salt conditions, an ionic strength generally around 0.15–0.2 M, and neutral pH. Once extracted, these conditions can be varied readily, though it is inconvenient to increase the protein concentration (by removing solvent), or to decrease the ionic strength substantially (by removing low-molecular-weight ions). The solubility is a result of polar interaction with the aqueous solvent, ionic interactions with the salts present, and to some extent repulsive electrostatic forces between like-charged molecules or small (soluble) aggre-

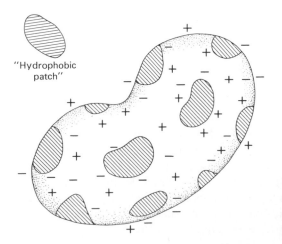

"Hydrophobic patch"

**Figure 3.1.** Distribution of charges and hydrophobic patches over surface of a typical protein.

gates of molecules. Small aggregates cancel out strong attraction charges (Figure 3.2). In the ionic strength range from zero to physiological, some proteins form precipitates because the repulsive forces are insufficient. For example, a high surface hydrophobicity means low interaction with the solvent and fewer charged groups to interact with salts. An overall charge near zero minimizes electrostatic repulsion, and may, close to the isoelectric point, result in electrostatic attraction of molecules to each other (Figure 3.3). This is called isoelectric precipitation; a globulin-type protein shows a minimum solubility around its isoelectric point, which may be low enough to result in precipitation. In a protein mixture the situation is greatly complicated by coprecipitation: different proteins with similar properties aggregate to form the isoelectric precipitate. In many cases isoelectric precipitates can be formed in a tissue extract by lowering the pH to between 6.0 and 5.0. This is not just an aggregation of soluble protein; it includes aggregates of particulate material such as ribosomes and membranous fragments, as discussed previously (cf. section 2.3).

The solubility of globulin-type proteins decreases as the salt concentration decreases. This can rarely be exploited with a crude extract since a twofold decrease in salt can only be achieved easily by adding an equal volume of water; if the solubility does not decrease by *more* than twofold, nothing is gained! However, it can be a useful property (though sometimes an inconvenience) at a later stage when desalting is attempted by dialysis or by gel filtration. On dialysis against a low-ionic-strength buffer, precipitates can form in

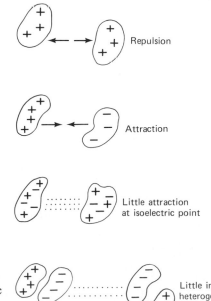

**Figure 3.2.** Electrostatic forces between protein molecules and small aggregates.

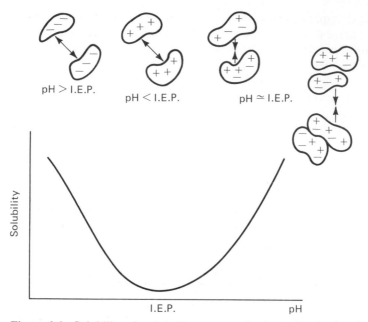

**Figure 3.3.** Solubility of a globulin-type protein close to its isoelectric point (IEP).

the dialysis sac. If the precipitate includes most of the enzyme required, or alternatively none at all, a purification results. The precipitate is simply centrifuged off. However, in gel filtration columns, development of an isoelectric precipitate can be a nuisance. Aggregates may clog up the column, or, precipitated and trapped, they remain behind until the salt removed from the original protein mixture catches up and redissolves the precipitate. Gel filtration desalting may simply not work under these conditions (cf. section 1.4).

Isoelectric precipitation of low-solubility proteins can be improved by inclusion of a solute which further decreases protein solubility. Thus organic solvents or polyethylene glycol (cf. sections 3.4 and 3.5) together with a pH adjustment can produce the desired precipitate. One example of such a method is in the purification of yeast phosphofructokinase (17).

The increase in solubility (at a given pH, temperature) with *increasing* salt concentration is known as "salting in." The opposite of this process, by dilution or dialysis, is more likely to be useful in enzyme purification. A typical salting-in curve for a pure protein is shown in Figure 3.4; the steepness of the response to salt may be sufficient to induce precipitation by dilution.

## Points to Note in Practice

Most isoelectric precipitates are aggregates of many different proteins and may include particulate fragments and protein–nucleic acid complexes. If the initial composition of the solution is changed, a desired enzyme may not exhibit the same apparent solubility behavior if its partners in precipitation are absent.

Since isoelectric precipitates normally occur by lowering the pH below physiological, make sure the enzyme is stable at the pH needed. Keeping the temperature low will improve the chances of stability and decrease solubilities. However, if adjusting the pH at 0°C, do not forget to standardize the pH meter at this temperature, using ice-cold standard buffer (cf. section 6.1).

In summary, the salting-in range can be exploited in purification in two distinct ways. First, aggregation may occur by dilution or dialysis, and if the required enzyme is present in the aggregate a useful purification will have been achieved. Second, isoelectric precipitation can be used, exploiting the variation of solubility with pH, without changing ionic strength. In either situation, if the required enzyme remains in solution, the degree of purification achieved is usually minimal since a relatively small proportion of the total protein is likely to be in the precipitate. In this case what one accomplishes is simply cleaning up the extract (cf. section 2.3).

## 3.3   Salting Out at High Salt Concentration

### General Aspects of Protein Solubility at High Salt Concentration

This section describes one of the most widely used techniques in enzyme purification, the salting out of proteins, using high concentrations of salts. It is quite distinct from the salting in effect described in section 3.2. Although there is a tendency for globulin-type proteins (those that have low solubility at low ionic strength near their isoelectric point) also to show relatively low solubility in high salt solutions, there are many exceptions. Thus serum γ-globulins, typical

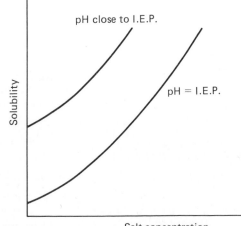

**Figure 3.4.** Solubility of globulin-type protein at a fixed pH close to its isoelectric point (IEP), varying with salt concentration.

proteins insoluble at low salt, also precipitate with ammonium sulfate at the relatively low concentration of around 1.5 M. On the other hand, some other serum globulins ($\alpha$- and $\beta$-) have fairly average ammonium sulfate solubilities, precipitating around 2.5–3 M. Salting out is largely dependent on the hydrophobicity of the protein, whereas salting in depends more on surface charge distribution and polar interactions with the solvent.

A typical protein molecule in solution has hydrophobic patches (side chains of Phe, Tyr, Trp, Leu, Ile, Met, Val). Forcing these into contact with the aqueous solvent causes an ordering of water molecules, effectively freezing them around the side chains (Figure 3.5) (25). This ordering is a thermodynamically unstable situation since it represents a large decrease in entropy compared with the unsolvated protein plus free water molecules.

If the process of dissolving a dry protein in water is represented by

$$P + nH_2O \rightarrow P \cdot nH_2O \tag{3.1}$$

where the $nH_2O$ represents ordered water around hydrophobic residues (we shall ignore the hydrogen-bonded and polar-bound water in this description), then $\Delta S^0$ is large and negative because of the degree of order introduced. The process is known to occur; thus $\Delta G^0$ is likely to be negative. Consequently $\Delta H^0$ is also highly negative ($\Delta H^0 = \Delta G^0 + T \Delta S^0$). Consider two protein molecules coming together and interacting through the hydrophobic residues, releasing the water:

$$P \cdot nH_2O + P' \cdot mH_2O \rightarrow P - P' + (n + m)H_2O \tag{3.2}$$

Clearly $\Delta S^0$ will be large and positive in this case; the enthalpy change is expected to be approximately the reverse of the process in Eq. (3.1), and so would also be positive. Now $\Delta G^0$ may be small, and its sign would depend on the relative magnitudes of $\Delta H^0$ and $\Delta S^0$. In low salt concentration, for a

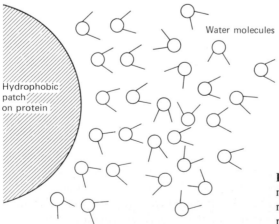

Water molecules

Hydrophobic patch on protein

**Figure 3.5.** Ordering of water molecules around hydrophobic residues on the surface of a protein.

typical water-soluble protein, $\Delta G^0$ would be positive. Since the protein is soluble, aggregation (Eq. 3.2) is not occurring to any significant extent. However, if the water molecules liberated by the process described in Eq. (3.2) are trapped, the reaction can be made to progress and protein aggregation will occur. Salt ions become hydrated; thus adding large amounts of salt traps the water:

$$AB + (n_1 + n_2)H_2O \rightarrow A^+ \cdot n_1H_2O + B^- \cdot n_2H_2O \qquad (3.3)$$
(salt)

Once a sufficient amount of water has been attracted away from the protein, the aggregation described by Eq. (3.2) occurs. Note that $\Delta H^0$ is positive for this reaction. As a result, the Arrhenius equation predicts that aggregation occurring due to hydrophobic forces increases with temperature. This is a recognized feature of hydrophobic interactions. Salt precipitation in most cases has a positive $\Delta H^0$, since more precipitate forms *at a given salt level* at higher temperatures.

Another way of describing the aggregation of hydrophobic patches on protein surfaces due to high salt concentration is to say that, as the salt ions become solvated, there is a greater tendency, when freely available water molecules become scarce, to pull off the ordered "frozen" water molecules from the hydrophobic side chains, so exposing the bare fatty areas which want to interact with one another. Those proteins with a larger number, or bigger clusters of such residues on their surface will aggregate sooner, whereas proteins with few nonpolar surface residues may remain in solution even at the highest salt concentration. Alternative interpretations of the effects of salt on protein solubility involve more complex considerations of free energy terms. The effectiveness of different salts has also been correlated with molal surface tension increment due to the salt being dissolved in the solvent (26).

Starting with a crude mixture, co-aggregation is extensive; individual identical molecules are not going to search each other out when another sticky molecule presents itself alongside. Nevertheless, many enzymes are precipitated from solution over a sufficiently narrow range of salt concentration to make the procedure a highly effective method of fractionation. Some results on the effects of the presence of other proteins on the solubility of some enzymes are shown in Figures 3.6 and 3.7. Pure protein solubility follows the empirical formula

$$\log(\text{solubility}) = A - m(\text{salt conc}) \qquad (3.4)$$

where $A$ = constant dependent on temperature and pH
$\quad\;\; m$ = constant independent of temperature and pH

A few examples of the values for pure protein are given in Table 3.1. Impure samples do not follow this rule, since coprecipitation occurs to an amount depending on the properties of the other protein present. In general, solubility of a particular component is decreased by the presence of other proteins.

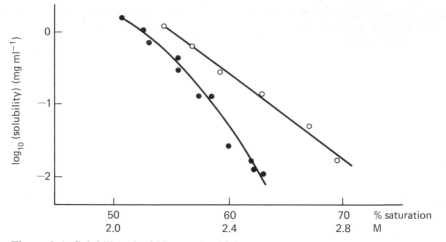

**Figure 3.6.** Solubility of rabbit muscle aldolase at pH 7.0 in ammonium sulfate solutions: ○—pure enzyme solubility curve; log(soly) $= 6.30 - 2.84$(salt conc); ●—solubility in a mixture of rabbit muscle enzymes: aldolase, pyruvate kinase, glyceraldehyde phosphate dehydrogenase, and lactate dehydrogenase in proportions 3:3:3:1. (Author's unpublished results.)

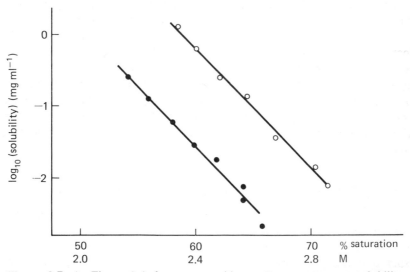

**Figure 3.7.** As Figure 3.6, for pyruvate kinase: ○—pure enzyme solubility curve; log(soly) $= 10.2 - 4.34$ (salt conc); ●—mixture of enzymes. (Author's unpublished results.)

Although salting out depends strongly on hydrophobic interactions, other features do affect solubilities. Thus pH may change the solubility; a removal of a charge makes the surface less polar. Solubility in salts is usually highest around pH 7, where proteins have the most charged groups. On the other hand,

**Table 3.1.** Some Values of the Constants $A$ and $m$ in Eq. (3.4) for Pure Enzymes at pH 7.0 (Solubility Expressed in mg ml$^{-1}$)

| Protein | $A$ | $m$ | Ammonium sulfate concentration for solubility of 1 mg ml$^{-1}$ (M) |
|---|---|---|---|
| Rabbit muscle aldolase | 6.30 | 2.84 | 2.20 |
| Rabbit muscle pyruvate kinase | 10.2 | 4.34 | 2.34 |
| Rabbit muscle glyceraldehyde phosphate dehydrogenase | 8.6 | 2.48 | 3.34 |
| Rabbit muscle lactate dehydrogenase | 8.0 | 3.97 | 2.02 |
| Rabbit muscle phosphoglycerate mutase | 9.85 | 4.26 | 2.32 |
| Bovine serum albumin | 9.2 | 3.26 | 2.84 |

*Source:* Author's unpublished observations.

aggregation may occur more easily close to the protein's isoelectric point. One classic example of this is in the one-step purification of glyceraldehyde phosphate dehydrogenase (EC 1.2.1.12) from rabbit muscle (27). At pH 6, this enzyme is soluble in 3.2 M ammonium sulfate, but if the pH is adjusted to 7.5 or higher, it precipitates, and may even crystallize, from the crude extract. The isoelectric point (at least at low salt concentration) is around 8.5.

Temperature affects the solubility of proteins at high salt concentration in an unusual way because of the effects of temperature on hydrophobic interaction as described above. In the salting-out range, the solubility of proteins generally decreases with increasing temperature. Many a time a turbid protein solution, prepared for crystallization at room temperature (cf. section 9.3), becomes completely clear when placed at 4°C.

## Practical Considerations in Salting Out Fractionations

Salting out of proteins involves dissolving the salt into the solution containing the proteins. Several features of this process should be considered. The nature of the salt is all-important. One must consider physical characteristics of the salt such as solubility, besides its effectiveness in causing precipitation. The most effective salts are those with multiple-charged anions such as sulfate, phosphate, and citrate; the cation is relatively less important. Salting-out ability of anions follows the Hofmeister series, which for some common anions is $SCN^-$, $I^-$, $ClO_4^-$, $NO_3^-$, $Br^-$, $Cl^-$, acetate$^-$, $SO_4^{2-}$, $PO_4^{3-}$. Although phosphate would appear from this to be more effective than sulfate, in practice phosphate at neutral pH consists of a mixture of $HPO_4^{2-}$ and $H_2PO_4^-$ ions, which are less effective than $PO_4^{3-}$. As far as cations are concerned, monovalent ions should be used, with $NH_4^+ > K^+ > Na^+$ in precipitation effectiveness. Of the common inexpensive salts that are effective in causing precipitation, sodium, potassium and ammonium sulfates, phosphates, and citrates are attractive candidates.

Further considerations of the suitability are the cost of kilogram quantities of sufficient pure salt, the solubility (several molar solutions are required), and the heat of solution. It is undesirable to get heating while dissolving in the salt; also, a large $\Delta H^0$ of solution means large changes in solubility with temperature. Finally, the density of concentrated solution should be considered, since centrifuging a precipitate depends on the difference in density between the aggregate and the solvent (cf. section 1.2).

Most potassium salts are ruled out on solubility grounds, though potassium phosphates can be used. Sodium citrate is very soluble, and can be used also, but rarely offers any advantage other than the possibility of use above pH 8—which is not possible with ammonium salts, because of the buffering action of ammonia. Citrate cannot be used so effectively below pH 7 for the same reason—the strong buffering action of citrate ions. Sodium sulfate forms several hydrates and has a complex solubility phase diagram, but it is not highly soluble at low temperature. It has been used in the purification of egg albumin, keeping the temperature around 30°C so that the salt is sufficiently soluble. Except when it is required to operate at a high pH, one salt has all the advantages and no disadvantages for a typical protein: ammonium sulfate.

The solubility of ammonium sulfate varies very little in the range 0–30°C; a saturated solution in pure water is approximately 4 M. The density of this saturated solution is 1.235 g cm$^{-3}$, compared with a protein aggregate density in this solution of about 1.29 g cm$^{-3}$ (cf. section 1.2). In contrast, 3 M potassium phosphate at pH 7.4 has a density of 1.33 g cm$^{-3}$, which makes it impossible to centrifuge off protein, since this is also the density of protein aggregates in this solution. However, filtration is possible, as is adsorption to amphiphilic matrices such as agarose (cf. section 4.8). Because of extra density caused by other salts and compounds present in crude extracts, precipitates in the range 3–4 M ammonium sulfate also can sometimes be difficult to sediment.

Ammonium sulfate is available in sufficiently pure state to be used in large quantities economically. However, small contaminating amounts of heavy metals, especially iron, could be detrimental to sensitive enzymes. Metal-complexing agents such as EDTA should be present in the solution before adding the ammonium salt. It is worth noting that 1 part per million of a heavy metal converts to the biochemically active concentration of a few $\mu$M, when the ammonium sulfate concentration is 3 M. The amount of ammonium sulfate to add to reach a predetermined concentration can be calculated from a simple formula (Eq. 3.5). Because the addition of salt causes an increase in volume, allowance for this is made in the formula. Thus, although the solubility of ammonium sulfate is 533 g liter$^{-1}$ at 20°C, 761 g must be added to 1 liter to make a saturated solution (1.425 liters). The density of the saturated solution is thus $1761/1425 = 1.235$ g cm$^{-3}$, as indicated above. It is usual to quote ammonium sulfate concentration as percent saturation, making the assumption that the starting material is equivalent to water in its ability to dissolve ammonium sulfate. For this, Eq. (3-5b) is used:

Grams ammonium sulfate to be added to 1 liter of a solution at $20°C$:

(a) At $M_1$ molar, to take it to $M_2$ molar:

$$g = \frac{533(M_2 - M_1)}{4.05 - 0.3M_2} \tag{3.5a}$$

(b) At $S_1\%$ saturation, to take it to $S_2\%$ saturation:

$$g = \frac{533(S_2 - S_1)}{100 - 0.3S_2} \tag{3.5b}$$

The two equations are related by the assumption that 100% saturation = 4.05 M. Note that in practice, because of other salts present and the protein itself, the *actual* solubility of ammonium sulfate in the sample will be somewhat less than 533 g liter$^{-1}$ at $20°C$. The amount of ammonium sulfate required can be read from Appendix A, Table I for 5% intervals; other intervals can be interpolated from these figures. Now pH may be important in precipitation, but in most cases small variations in pH from one experiment to another make little difference. It is best to operate at a neutral value (6–7.5); ammonium sulfate has a slight acidifying action, so around 50 mM buffer (e.g., phosphate) should be present.

One of the best-studied systems for ammonium sulfate fractionation is rabbit muscle extract. Czok and Bücher (27) divided the extract into 5 fractions by ammonium sulfate precipitation; in each fraction certain enzymes were especially enriched. Using this work as a basis, all glycolytic enzymes plus a few others could be purified after an initial division into 4 ammonium sulfate fractions (28).

One great advantage of ammonium sulfate fractionation over virtually all other techniques is the stabilization of proteins that occurs. A 2–3 M ammonium sulfate suspension of protein precipitate or crystals is often stable for years, and it is the normal packaging method for commercial enzymes. The high salt concentration also prevents proteolysis and bacterial action (cf. section 6.2). It is always useful to have a stage in a purification where the sample can be left overnight. An ammonium sulfate precipitate, either before centrifuging or as a pellet, is a good form to keep a sample in—although in some cases freezing might be better.

After deciding to do an ammonium sulfate fractionation, the next decision is what percentage saturation to try. Material precipitating before 25% saturation is generally particulate and preaggregated or very-high-molecular-weight protein. A typical distribution of protein precipitation from a crude extract is shown in Figure 3.8. An idealized set of results of trial fractionations with ammonium sulfate is reproduced in Table 3.2. Fractionations are always a compromise between recovery of activity and degree of purification. If the source material is valuable or difficult to obtain, or the objective is to get as

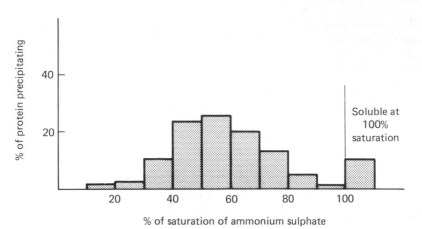

**Figure 3.8.** Typical ammonium sulfate precipitation ranges for proteins from a crude tissue extract.

**Table 3.2.** Trial Fractionations with Ammonium Sulfate

|              | Percent saturation range | Percent enzyme precipitated | Percent protein precipitated | Purification factor[a] |
|--------------|--------------------------|-----------------------------|------------------------------|------------------------|
| First trial  | 0–40                     | 4                           | 25                           |                        |
|              | 40–60                    | 62                          | 22                           | 2.8                    |
|              | 60–80                    | 32                          | 32                           | 1.0                    |
|              | 80 supernatant           | 2                           | 21                           |                        |

Conclusion: Enzyme precipitated more in 40–60% than in 60–80%; try 45–70%:

| Second trial | 0–45            | 6  | 32 |     |
|--------------|-----------------|----|----|-----|
|              | 45–70           | 90 | 38 | 2.4 |
|              | 70 supernatant  | 4  | 30 |     |

Conclusion: Good recovery, but purification factor not as good as in first trial; if purity important, try 48–65%:

| Third trial | 0–48            | 10 | 35 |     |
|-------------|-----------------|----|----|-----|
|             | 48–65           | 75 | 25 | 3.0 |
|             | 65 supernatant  | 15 | 40 |     |

[a]Specific activity of fraction divided by specific activity of original sample.
*Source:* Scopes (29).

much enzyme as possible without too much concern for absolute purity, then degree of purification may be sacrificed for recovery. On the other hand, if the fractionation is the first step from a readily available source material, recovery can be sacrificed for purity. In many cases more sophisticated processes [ion exchange chromatography (cf. section 4.3), affinity adsorption chromatography (cf. section 4.5)] may follow the ammonium sulfate fractionation and the behavior of the desired enzyme may be similar regardless of the amount of impurity. In that case broad ammonium sulfate cuts can be used, resulting in relatively little purification, the step being essentially a cleaning-up operation to get a messy extract into a suitable state for adsorption chromatography.

The solution to which salt is to be added should be provided with a stirring system, and solid ammonium sulfate is weighed out. Any lumps should be broken up before slowly adding the solid. The first bit can be dissolved in quite quickly, but as the intended percent saturation is approached the last increments of salt should be added more slowly. Dissolved air may come out of solution and cause frothing, but no harm is done unless overvigorous stirring itself causes frothing, in which case surface tension effects on proteins caught in bubbles can cause denaturation. After the last bit of salt has dissolved, stirring should continue for 10–30 min to allow complete equilibration between dissolved and aggregated proteins; then the solution is centrifuged. Generally around $10^5$ $g$/min is sufficient, i.e., 10,000 $\times$ $g$ for 10 min or 3000 $\times$ $g$ for 30 min, although at higher salt concentrations somewhat more centrifugation may be needed. The supernatant is decanted, its volume noted, and the amount of salt required for the next cut calculated. The precipitate is dissolved in a suitable buffer. Note that the volume of buffer need be no more than 1–2 times the volume of the *precipitate,* since that will be enough to reduce the salt concentration to well below the precipitation point of the proteins present. If all the precipitate does not dissolve, the insoluble material is probably denatured/particulate, and should be removed by centrifugation. It is always best to keep solutions as concentrated in protein as possible. This improves stability, and it is always easier to dilute something later than it is to reconcentrate it.

The redissolved precipitate contains considerable amounts of ammonium sulfate, and usually this must be removed before proceeding to the next step. Dialysis or gel filtration on a "desalting" column are the normal methods employed (cf. section 1.4). Measurement of enzyme activity can be carried out on the redissolved precipitate or on the supernatant, but beware that in the latter case the amount of ammonium sulfate added into the assay mixture with the sample could upset the assay; the precipitate should always be tested anyway, for although losses in ammonium sulfate fractionation are usually small, the activity recovered in a precipitate may not be as much as the difference between successive supernatants.

Rather than dissolve solid salt into the sample, it is possible to add a suitable volume of saturated solution, rather in the same way as adding organic solvents (cf. section 3.4). This is particularly useful if the enzyme is all precipitated by 50% saturation, as this would only involve a twofold increase in volume. On a large scale this would be unacceptable, but on a small scale it can be an appropriate method. Higher percent saturation needs still more increase in volume, and dilution of the proteins leads to the requirement of a still higher percent saturation for equivalent precipitation, so the method is not advised for going much above 50%.

An important point to remember is that salt never precipitates *all* the enzyme, but just reduces its solubility. If the starting mixture contains your enzyme at, say, 1 mg ml$^{-1}$, then a reduction in its solubility to 0.1 mg ml$^{-1}$ causes 90% of it to precipitate. This would be an acceptable recovery. On the other hand, if your enzyme was only at 0.1 mg ml$^{-1}$ to start with, the same concentration of salt would not cause any precipitation! In other words, the salt concentration range for "precipitation" is not an absolute property of the enzyme concerned, but depends on both the properties of other proteins present (coprecipitation) and on the protein concentration in the starting solution. Different salt concentrations may be needed at different stages in the purification procedure.

Finally, an important application of salting-out procedures is not so much for fractionating samples, but for concentrating them. An eluate from an ion exchange column or from gel filtration may need concentrating. Salting out by adding sufficient ammonium sulfate to precipitate all the protein is an efficient way of doing this—provided that the sample is not too dilute to start with. It is convenient to dissolve in 60 g ammonium sulfate for every 100 ml, to give an approximately 85% saturated solution. Few proteins have solubilities in excess of 0.1 mg ml$^{-1}$ at this level of ammonium sulfate. But if the starting protein concentration is less than 1 mg ml$^{-1}$, one should try to increase this by, e.g., ultrafiltration (cf. section 1.4) before salting out.

## 3.4   Precipitation with Organic Solvents

### General Theory

The method of protein precipitation by water-miscible organic solvents has been employed since the early days of protein purification. It has been especially important on the industrial scale, in particular in plasma protein fractionation (30). However, its use in laboratory-scale purifications has been less extensive; indeed it has been under-utilized, considering there are some advantages compared with salting-out precipitation; but there are also disadvantages. As the principles causing precipitation are different, it is not necessarily an alternative to ammonium sulfate, but can be used as an additional step.

Addition of a solvent such as ethanol or acetone to an aqueous extract containing proteins has a variety of effects which, combined, lead to protein precipitation. The principal effect is the reduction in water activity. The solvating power of water for a charged, hydrophilic enzyme molecule is decreased as the concentration of organic solvent increases. This can be described in terms of a reduction of the dielectric constant of the solvent, or simply in terms of a bulk displacement of water, plus the partial immobilization of water molecules through hydration of the (organic) solvent. Ordered water structure around hydrophobic areas on the proteins' surface can be displaced by organic solvent molecules, leading to a relatively higher "solubility" of these areas. Some extremely hydrophobic proteins, normally located in membranes, may be soluble in 100% organic solvent. But the net effect on cytoplasmic and other water-soluble proteins is a decrease in solubility to the point of aggregation and precipitation.

The principle causes of aggregation are likely to be electrostatic and dipolar van der Waals forces, similar to those occurring in the salting-in range in the absence of organic solvent. Hydrophobic attractions are less involved because of the solubilizing influence of the organic solvent on these areas. It has been found that precipitation occurs at a lower organic solvent concentration at around the isoelectric points of the proteins, which supports the suggestion that the aggregation is similar to that occurring in isoelectric precipitation. A two-dimensional representation of proteins in a water/organic solvent mixture is shown in Figure 3.9. Here aggregation is occurring by interactions between opposite-charged areas on the proteins' surfaces.

Another feature affecting organic solvent precipitation is the size of the molecule. Other things being equal, the larger the molecule, the lower percentage of organic solvent required to precipitate it. Thus, if one could compare a range of proteins with different molecular sizes but similar hydrophobicity and isoelectric points, the precipitation order would be the reverse of the order of size. In practice the different compositions and net charges of proteins in a given mixture make this an approximate rule at best. Large molecules aggregate sooner because they have a greater chance of possessing a charged surface area that matches up with another protein (Figure 3.9). Some experimental results

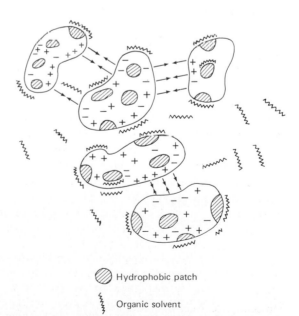

**Figure 3.9.** Aggregation of proteins by interactions in an aqueous–organic solvent mixture.

⬯ Hydrophobic patch

〜 Organic solvent

showing both the size relationship and the isoelectric point effect have been presented previously (29) and are reproduced in Figures 3.10 and 3.11. In an early investigation of organic solvent fractionation, it was concluded that acetone, at pH 6.5, was the most suitable for rabbit muscle extracts (31). The isoelectric points of the major components of mammalian muscle extracts, in phosphate buffer, are in the range 6.0–7.5.

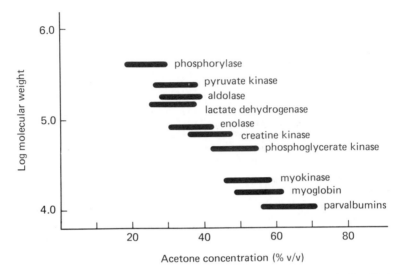

**Figure 3.10.** Approximate precipitation ranges in acetone at $0°C$, pH 6.5, $I = 0.1$, of some proteins found in muscle tissue extracts. [From Scopes (29).]

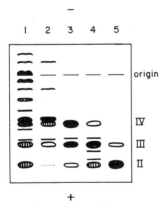

**Figure 3.11.** Starch gel electrophoretogram of acetone fractions of a carp muscle extract, pH 6.5, showing the separation of parvalbumins (MW approx. 11,000) II, III and IV. (1) whole extract; (2) acetone fraction 55–60% v/v; (3) 60–65%; (4) 65–70%; (5) 70–80%. The most acidic band was precipitated last, because it was the furthest removed from its isoelectric point (approx. 4.0). [From Scopes (29).]

## Choice of Solvent

The solvent used must be completely water-miscible, unreacting with proteins, and have a good precipitating effect. The two most widely used solvents are ethanol and acetone. Others that can be used include methanol, *n*-propanol, *i*-propanol, dioxan, 2-methoxyethanol (methyl cellosolve), and various other more exotic alcohols, ethers, and ketones. Safety should be a major consideration, both in terms of flammability and in terms of noxious vapors. For these reasons dioxan, 2-methoxyethanol and other ethers should be eliminated from the list.

One advantage of organic solvent fractionation is that it can be carried out at subzero temperatures, since all the miscible solvents form mixtures with water that freeze well below $0°C$. This is fortunate, since it is most important that the temperature is kept low. At above about $+10°C$ denaturation effects become substantial. The reason for denaturation concerns the intramolecular hydrophobic interactions which help maintain protein structure. At low temperatures, the lack of conformational flexibility means that organic solvent molecules are unlikely to penetrate the internal structure and cause destabilization. But at higher temperatures small organic molecules enter "cracks" in the surface which occur spontaneously due to natural flexing of the structure and attach themselves through hydrophobic forces to internal residues such as Leu, Ile, Tyr, Phe, Val, etc. (Figure 3.12). At higher temperatures the internal hydrophobic forces in the protein molecule are stronger, and relatively more important in maintaining the molecule's integrity; loss of these interactions quickly results in autocatalytic denaturation.

Certain organic solvents are more effective at denaturing proteins than others. It has long been recognized that longer-chain alcohols are more denaturing than short-chain ones. An illustration of this is presented in section 3.5. Although ethanol is used in standard procedures for plasma protein fractionation (30), plasma proteins as a group are unusually stable—few are enzymes—and they survive conditions which cause denaturation of more sensitive proteins. Acetone has a lesser tendency to cause denaturation than ethanol, partly because slightly lower concentrations are needed to cause comparable amounts of precipitation at low temperatures. It is also more volatile, which enables it to be removed easily from redissolved precipitates under reduced pressure.

## Operating Procedures

Because the temperature must be kept low, an ice-salt bath or other container that can be maintained below $0°C$ should be prepared and the sample undergoing fractionation chilled to $0°C$. The protein concentration can be $5-30$ mg $ml^{-1}$, but the salt concentration should not be too high. If the salt concentration is high (e.g., subsequent to an ammonium sulfate fractionation), then electrostatic aggregation is impaired, higher levels of organic solvent are needed, and denaturation is more likely. On the other hand, at very low salt concentrations

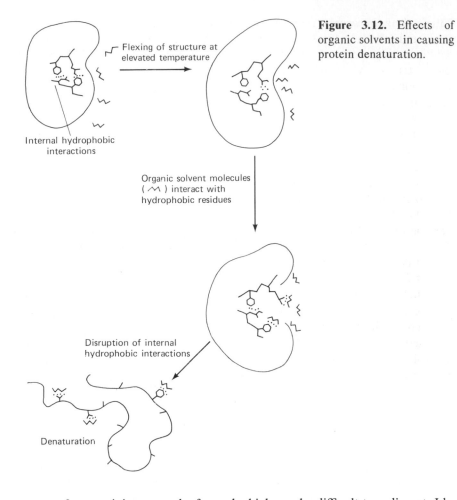

**Figure 3.12.** Effects of organic solvents in causing protein denaturation.

Flexing of structure at elevated temperature

Internal hydrophobic interactions

Organic solvent molecules ( ∿ ) interact with hydrophobic residues

Disruption of internal hydrophobic interactions

Denaturation

a very fine precipitate may be formed which can be difficult to sediment. Ideally the starting value of the ionic strength should be 0.05–0.2. Addition of the solvent to water causes heat evolution due to the negative $\Delta H^0$ of hydration of solvent molecules. Consequently slow addition with efficient cooling is the procedure to follow (use a glass container, not plastic, for good thermal conductivity; stainless steel beakers are even better). If the enzyme proves to be particularly unstable to organic solvent treatment, it may be best to allow the temperature to fall below 0°C almost at once, as the added solvent lowers the freezing point. But most enzymes are much more resistant than this, and it is probably unnecessary to maintain the temperature much below 10°C while the solvent remains below 20% v/v. After this amount of solvent has been added, no further heating occurs, so a constant temperature can be maintained with ease. A typical temperature curve for continuous addition of (cold) organic solvent, with efficient cooling, is shown in Figure 3.13.

When the first cut, usually between 20 and 30% (acetone or ethanol), is taken, the temperature must be controlled closely at the desired value, e.g., 0°C, and left there for 10–15 min before centrifuging in a precooled rotor to maintain that temperature. The amount of precipitation is highly dependent on the temperature and is greater at lower temperature values. (This is a rare exception to one of nature's frustrating laws which states that a change of one thing makes another one worse. Decreasing temperature will not only lessen the chance of denaturation in itself, but it will also allow the use of a lower level of organic solvent.) Since subzero temperatures are difficult to maintain accurately, especially during centrifugation, a standard practice is to use "0°C ± 1°C" for organic solvent precipitations.

Most enzymes precipitate with acetone in the range 20–50% v/v. There is a problem in defining percentages by volume; properly quoted, a percentage should add the statement "assuming additive volumes." For if one adds 50 ml acetone to 50 ml water, the result is only 95 ml of liquid. To describe this as "50% v/v acetone assuming additive volumes" is clearly not correct, but at least it defines the operation. Similarly, a 50:50 mixture of ethanol and water loses 4% of the additive volume. The volume loss is due to the formation of hydrated solvent complexes which occupy a smaller volume than their constituent components. The volume change introduces a marginal difference between reaching an assumed percentage in one step, e.g., 0–50% by adding an equal volume of solvent, and reaching it stepwise; but this is unlikely to be of significance. Unless clearly indicated otherwise in an experimental protocol, assume that quoted percentages are "v/v assuming additive volume."

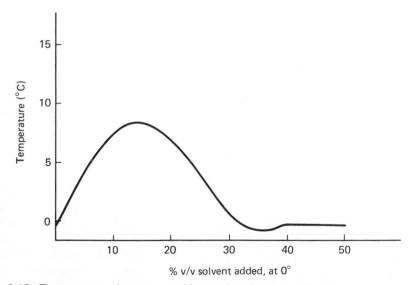

% v/v solvent added, at 0°

**Figure 3.13.** Temperature changes as cold organic solvent is added to an aqueous solution, with efficient cooling. After about 10% solvent is reached, the heating effect of hydrating more solvent is less than the external cooling. After 20% solvent there is little or no more heating effect.

After equilibrating the mixture at the given percentage solvent at a fixed temperature for 10–15 min, the precipitate can be centrifuged. Because most miscible organic solvents have lower densities than water, the sedimentation of aggregated protein can be very rapid. In fact, at high percentages, when volumes may also be large, sedimentation at 1 $g$ (i.e., on standing) may be sufficiently fast to allow decantation of much of the supernatant without using a centrifuge. But centrifugation is usually necessary, and large-capacity centrifuges capable of generating only a few thousand $g$ are quite satisfactory, provided that they are refrigerated. For reproducible results the need for constant temperatures both before and during centrifugation cannot be stressed too much.

The precipitate can now be redissolved, using *cold* buffer of suitable composition. If a precipitate is not dispersing to make a clear solution, do not add more and more buffer—there is probably denatured protein present. It is usually easier to disperse protein precipitates in a small volume. Doubling the volume of the precipitate will halve the organic solvent concentration, which should allow solution of all undenatured protein. Once a solution has been excessively diluted it is difficult to concentrate again. Moreover, if excess organic solvent is to be removed by evaporation, the higher the starting concentration the more quickly solvent will be lost.

Small amounts of organic solvent, e.g., up to 10% v/v, are unlikely to affect other fractionation methods, with the exception of hydrophobic chromatography (cf. section 4.8), affinity adsorptions which depend on hydrophobic interactions (cf. section 4.5), and often salting out fractionation. Provided that the enzyme is relatively stable toward the solvent, ammonium sulfate fractionation can be carried out in the presence of the solvent, though the salt concentration required for precipitation is likely to be somewhat higher due to organic solvent molecules interfering with the hydrophobic aggregation at high salt concentration. Excessive amounts of solvent can be removed by placing the solution in a Büchner flask and evaporating at reduced pressure. The temperature may be allowed to rise to 25–30°C by immersing the flask in warm water during this process. However, some solutions froth excessively, and this method is not always satisfactory. The best way to remove all trace of organic solvent, plus any other low-molecular-weight compounds that may have precipitated with the protein, is to use a gel filtration column appropriate to the volume of sample (Table 1.2).

To the supernatant from the first organic solvent precipitation, more solvent is added to reach the next percentage value. A formula for calculating the amount of organic solvent to be added is

$$\text{Volume to add to 1 liter to take \% from } x \text{ to } y = \frac{1000(y - x)}{100 - y} \text{ ml}$$

A table of amount of pure solvent to add at each level is presented in Appendix A (Table II).

At 50% solvent only proteins of molecular weight less than 15,000 are likely to remain in solution. Precipitation of acidic proteins may be aided by inclusion of divalent metal ions such as magnesium, which form magnesium protein complexes with reduced net charge, encouraging aggregation. Alternatively, a pH change toward the isoelectric point may result in precipitation without increasing the organic solvent concentration. Note that pH values, as measured directly with a pH electrode standardized against aqueous buffer, will generally increase as organic solvent content increases. A simple way of explaining this is to consider an acid dissociation to be the sum of two processes:

$$HA = H^+ + A^-$$
$$H^+ + H_2O = H_3O^+$$

The second process is affected because the water activity is reduced by organic solvent; thus $K_a$ for the overall process is smaller; $pK_a$ is larger. As an example, a recent preparation of yeast phosphofructokinase (17) involved adjusting the extract to pH 6.0, then adding acetone to 23% v/v and removing the precipitate. The pH of the supernatant then read 6.4.

It has been stressed that denaturation can occur if the temperature is too high in organic solvent fractionation. But denaturation is selective; some enzymes are remarkably stable. In particular extracellular proteins and enzymes from thermophilic organisms have to be more robust because of their natural environment. Such proteins often survive organic solvents at elevated temperatures. Use of organic solvents for selective denaturation is discussed further in section 3.6.

## 3.5 Precipitation with Organic Polymers and Other Materials

Salts and organic solvents are not the only materials which can cause aggregation of proteins without denaturation. Polson et al. (32) investigated the ability of a variety of high-molecular-weight, neutral, water-soluble polymers to precipitate plasma proteins. Although several were effective in causing precipitation, the high viscosity of most of the solutions made their use as protein precipitants impractical. There was one exception, polyethylene glycol, which is available in a variety of degrees of polymerization. Solutions of this polymer up to 20% w/v are not too viscous, and many components of the plasma precipitate before 20% w/v is reached. Polyethylene glycol of molecular weight 4000 or greater is most effective; the two types commonly used for protein precipitation have molecular weights of about 6000 and 20,000. The behavior of the proteins is rather similar to their behavior in precipitation by organic solvents, and indeed the polyethylene glycol (PEG) molecule can be regarded as polymerized organic solvent, although the percentage required to cause a given amount of precipitation is lower. The first plasma protein precipitated by PEG is fibrinogen, a rather large, highly asymmetric molecule which is only just "soluble" in the plasma in the first place. Then (at neutral pH close to their isoelectric points) γ-globulins precipitate, followed by various other components. As with organic solvents, proteins become more soluble in PEG solutions as the pH moves away from their isoelectric points.

PEG precipitation has been used with considerable success for intrinsically low-solubility proteins, such as "globulins." For other "albumins," such a high concentration of PEG may be needed to cause a precipitate that (a) the viscosity and density of the solution make problems in centrifuging and (b) in complex protein extracts, phase separation can occur, with a protein-rich heavy phase separating from a lighter phase above.

PEG is not as easy to remove from a protein fraction as either salt or organic solvent; as a polymer it will not dialyse rapidly, and separation on conventional Sephadex G-25 columns may not be good—especially if MW 20,000 PEG was used. Nevertheless, a residual low level of PEG is not detrimental to many procedures; salting out, ion exchange, affinity chromatography, or gel filtration can be carried out without having to remove the PEG first.

Charged polymers (polyelectrolytes) have also been used for protein purification (30,33), especially in industrial applications, where their cheapness and lack of waste disposal problems make them attractive. This is because precipitation occurs at very low percentage, around 0.05 to 0.1 w/v. But their use is restricted to specific cases because of the mode of precipitation; electrostatic complexes form between polyacrylic acid salts and positively charged proteins at somewhat low pH values, usually in the range 3–6; many enzymes do not withstand these low pH values. Moreover, many enzymes are not sufficiently positively charged until the pH is at the lower end of this range. Sternberg and Hershberg (33) described the precipitation of various proteins by both linear and cross-linked polyacrylates. Basic proteins such as lysozyme, cytochromes, protamines, and trypsin were precipitated effectively, though pepsin, negatively charged at pH 3.5, was also partly precipitated at this pH. It is possible that positively charged polymers such as DEAE-dextran might be effective for acidic proteins in a higher pH range.

Other precipitants include caprylic acid salts (n-octanoic acid) and rivanol (2-ethoxy-6,9-diamino acridine lactate), principally in plasma protein fractionation (30). It is probable that a wide variety of other organic compounds possessing dual functions of hydrophobicity and polarity could cause precipitation, but such techniques have not been extensively investigated.

## 3.6 Precipitation by Selective Denaturation

### General Principles

Almost every other section of this book stresses treating enzymes by gentle procedures to avoid denaturation. Yet there are a large number of enzymes which are very robust and stand quite extreme conditions. One can use this exceptional stability by exposing an impure preparation to these conditions, whereby unwanted proteins denature and precipitate out of solution. Denaturation implies destruction of the tertiary structure of a protein molecule and formation of random polypeptide chains. In solution these get tangled together and aggregate physically, and to some extent chemically through formation of $-S-S-$ bonds. Nevertheless, denatured proteins in low-salt conditions well away from their isoelectric point may remain completely in solution and only precipitate when the pH is adjusted. This is because repulsion between the charged random polypeptides keeps them apart; closer to their isoelectric point they can aggregate, or higher salt concentration can disrupt the intermolecular repulsive forces and allow aggregation.

The objective of a selective denaturation step is to create conditions where the enzyme in question will not quite denature—or at least only up to 10–20% is lost, whereas many other components are denatured, and preferably precipitated, by the treatment. The main stresses imposed are (1) temperature, (2) pH, and (3) organic solvents. These three are not generally independent since temperature denaturation depends strongly on pH and vice versa, and with organic solvents, temperature, pH, and ionic strength must be carefully defined. Ionic strength may make some difference to (1) and (2) also, especially if comparing behavior at low $I$ ($\sim 0.01 \rightarrow 0.1$) with that at high salt concentration, e.g., M ammonium sulfate. The three types of denaturation are described separately below.

## Temperature Denaturation

Denaturation of proteins by heat can be described as a first-order process with the rate constant obeying normal thermodynamic principles. Thus the rate constant $k_{den}$ can be related to temperature by

$$\frac{d \ln k_{den}}{dT} = \frac{E_{act}}{RT^2}$$

where $E_{act}$ = energy of activation = $\Delta H^{\ddagger} + RT$
 $\Delta H^{\ddagger}$ = enthalpy of activation

What makes heat denaturation of proteins so different from normal chemical processes is the high value of $E_{act}$ ($\Delta H^{\ddagger}$) in the expression above; consequently the rate of denaturation changes very rapidly with temperature. According to absolute reaction rate theory $\Delta G^{\ddagger}$ can be calculated, and for denaturation rates in the range $10^{-4}$ to $10^{-2}$ sec$^{-1}$, it is around 80 kJ mol$^{-1}$. Since $\Delta H^{\ddagger} = \Delta G^{\ddagger} + T \Delta S^{\ddagger}$ and $\Delta H^{\ddagger}$ is typically from 300–500 kJ mol$^{-1}$, it is seen that $\Delta S^{\ddagger}$ is an unusually high figure, between 200 and 1000 J deg$^{-1}$ mol$^{-1}$. The explanation is simple; initiation of denaturation involves the rupturing of many bonds, a marked conformational change in the protein, and release of associated solvent molecules, which is accompanied by an increase in entropy—a large amount of disordering. Now $\Delta S^{0}$ for the total denaturation process, the complete randomization of polypeptide chains, can be several thousand J deg$^{-1}$ mol$^{-1}$. The whole process of denaturation can involve several steps along many alternative routes, but for the thermodynamic description of the overall process, only the activation energy of the slowest step needs to be known, and this is usually the first step.

The theoretical curve for amount of protein denatured at a given temperature in 10 min is shown in Figure 3.14, where it has been assumed that $E_{act} = 400$ kJ mol$^{-1}$ and that 50% denaturation in 10 min occurred at 50°C. Note that at only 5°C above this half-denaturation temperature, barely 1% activity remains, yet at 5°C below, as much as 8% activity is lost.

Different proteins will have differing $E_{act}$, and so the midpoints of their denaturation curves will vary widely (as will the shapes). Consequently it is possible to choose a temperature which completely denatures one protein, while leaving over 95% of another protein unaffected; a collection of proteins might have individual denaturation curves, as illustrated in Figure 3.15. In this case, if one were attempting to purify protein D, heating for 10 min at 50°C would result in almost total denaturation of A and B, and a little of C. But 90% of D would remain, and all of E and F. Depending on the relative amounts of these proteins in the mixture being treated, this might be a worthwhile step, especially if it was otherwise difficult to separate A or B from D.

As with so many other empirical fractionation methods, it is highly unlikely that one would have any idea of the behavior of many, if any, of the proteins in the mixture—certainly nothing like the detail of Figure 3.15. But the behavior of the particular enzyme of interest can be determined by using small-scale trials at, say, 5°C temperature intervals. The time of incubation is only important for reproducibility; 1 min or 60 min will give the same *shapes* of curves, shifted along the temperature axis. But whereas 1 min might be very convenient on a 1-ml scale, it would be almost impossible to heat up, then cool rap-

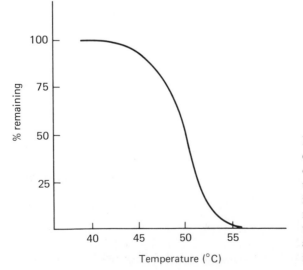

**Figure 3.14.** Theoretical diagram showing the percentage of protein remaining undenatured after 10 min incubation. $E_{act} = 400$ kJ mol$^{-1}$, and 50% denaturation occurred at 50°C. It is assumed that the process is irreversible.

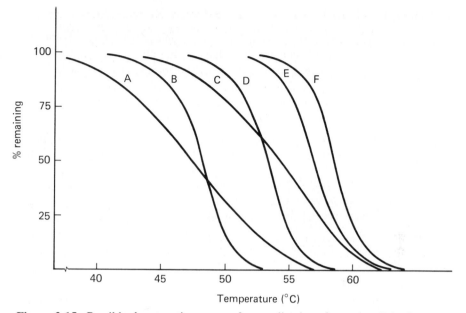

**Figure 3.15.** Possible denaturation curves for a collection of proteins; F is the most stable to heat; A and B the least stable (see text).

idly a large volume within a short time. While approaching and leaving the incubation temperature, further denaturation will occur.

An important consideration with heat denaturation is the possibility of proteolysis. Proteases, like other enzymes, are more active at higher temperatures. They tend to be fairly stable, so even if the desired enzyme remains active, there is a chance of nicking or minor modification spoiling the final preparation. For this reason it is preferable to do a heat step in the presence of ammonium sulfate (despite the fact that the salt may stabilize the proteins further, so that higher temperatures are required); proteolysis is largely inhibited by the high salt concentration. There are also documented examples where the presence of substrate further stabilizes its particular enzyme; however, destabilization by substrate is also possible.

The buffer composition of the sample to be heat-treated must be adjusted carefully in order to get reproducible results; the temperature profile must be reproducible, and above all the pH of the solution must be carefully defined. As indicated in the subsection below, a very small difference in pH can make a large difference to the amount of denaturation.

Heat denaturation trials can be made in a range of pH values; what one is looking for is the maximum loss (precipitation) of unwanted proteins at a given recovery, say 90%, of the enzyme. At the extremes of pH, temperatures will need to be lower, and this now is more properly described as pH denaturation.

## pH Denaturation

The further one moves away from the physiological operating pH of the enzyme, the less stable it becomes. Note that it is not always a matter of being away from neutrality; pepsin is very stable and normally active at pH 1–2, but denatures quickly at pH 7 or higher. Nor is it necessarily true that the optimum stability coincides with the optimum pH of an enzyme—the optimum pH is frequently not relevant physiologically, since activity under experimental conditions of substrate saturation does not necessarily correspond to physiological conditions. Nevertheless, it is expected than an enzyme will be maximally stable at or close to its physiological environmental pH.

Extremes of pH cause denaturation because sensitive areas of the protein molecule acquire more like charges, causing internal repulsion, or perhaps lose charges which were previously involved in attractive forces holding the protein together (Figure 3.16). The most stable pH is not necessarily at the isoelectric point; indeed, many bacterial proteins are isoelectric in the pH range 4–5 but unstable in that range; they exist at a pH of 6–7 in the cell. But the susceptible *areas* illustrated in Figure 3.16 might be effectively "isoelectric" at the physiological pH.

The rate of denaturation in acid/alkaline conditions is not the protonation step illustrated in Figure 3.16: protonation/deprotonation of charged residues is virtually instantaneous. It is the opening-up step which determines the rate, and this process is essentially the same as occurring in "heat" denaturation.

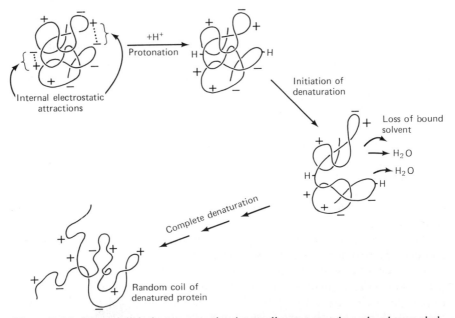

**Figure 3.16.** Electrostatic forces operating internally on a protein molecule are abolished by protonation, leading to opening up of molecule and complete denaturation.

Thus temperature is all-important, and in these more extreme pH conditions temperatures around 0–10°C rather than 50–60°C may be denaturating. In view of the great difference that a few degrees makes, pH denaturation should be carried out at a carefully specified temperature. Now pH adjustment can be made quickly down to the desired value, and up again (or up and down again), so there is no problem of slow approaches to conditions when scaling up the procedure. Also, pH adjustment should be carried out with an appropriate acid or base—this is discussed further in section 6.1. Strong acids and bases should be avoided unless necessary. Tris and acetic acid can be used for adjusting pH values in the range 5–8.5; stronger acids or bases are needed only to go outside this range. Lactic acid is suitable down to about pH 4.0, after which phosphoric or even sulfuric acid might be needed. If the enzyme is stable at pH 2, then brief exposure to a drop of strong acid is less likely to cause harm to it than to a protein which denatures at about pH 5. For high pH, diethanolamine (to pH 9) or sodium carbonate (to pH 10.5) can be used; sodium or potassium hydroxide will be needed to get above pH 11. In these extremes of pH it is likely that a substantial proportion of denatured proteins will remain in solution, even if the salt concentration is moderate. But on readjusting to neutrality, denatured proteins will normally precipitate out. It is advisable, after neutralization, to incubate briefly at 25–30°C to encourage completion of the aggregation of the denatured proteins before centrifuging off the precipitate.

## Denaturation by Organic Solvents

In section 3.4 the use of organic solvents as protein precipitants at low temperatures was described. The temperature is maintained around 0°C to avoid denaturation. In this section the use of organic solvents as denaturants is described; the temperature is usually kept at 20–30°C or even higher to encourage the process, and the concentration of organic solvents used is usually lower than that needed to cause precipitation of native proteins.

The principle rests on the differential sensitivity of proteins to this treatment. Some proteins are remarkably stable, and this property should be exploited, allowing other, less stable proteins to be denatured. As examples, rabbit muscle creatine kinase is stable in alkaline conditions to 60% v/v ethanol at 25°C, and this was used as a step in the original purification of this enzyme (34). Yeast alcohol dehydrogenase can be purified easily after an initial treatment with 33% v/v ethanol at 25°C (35).

An unpublished study of the effect of alcohols was carried out on yeast glyceraldehyde phosphate dehydrogenase. Using different alcohols it was found that the longer the aliphatic chain the more denaturing the alcohol was (Figure 3.17). The percentage of n-alcohol required to cause 50% denaturation in 30 min at 30°C decreased by a factor of *exactly* 2 for every methylene added to the alcohol. Thus the percentages for methanol, ethanol, propan-1-ol, butan-1-ol and pentan-1-ol were 34, 17, 8.5, 4.2, and 2.2 respectively; lack of miscibility

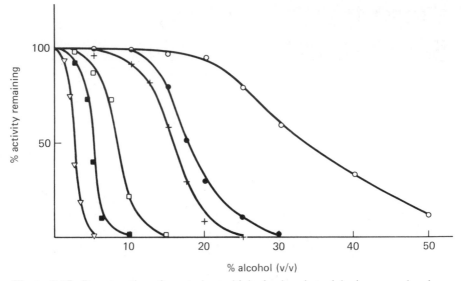

**Figure 3.17.** Denaturation of yeast glyceraldehyde phosphate dehydrogenase by alcohols. Enzyme at 2 mg ml$^{-1}$ was incubated at 30°C for 30 min, cooled, and any precipitate removed by centrifugation. A sample of the supernatant was assayed for enzyme activity remaining. ○—methanol; ●—ethanol; □—propan-1-ol; ■—butan-1-ol; ▽—pentan-1-ol; +—propan-2-ol. (Author's unpublished results.)

of longer-chain alcohols prevented this from being extended further. Branched chain alcohols were less denaturing (and more water-miscible), indicating that the length of aliphatic chain is the important feature. Acetone gave very similar results to ethanol. It would appear from these results that it might be safer to use methanol for organic solvent precipitation if denaturation is a problem.

The pH and temperature must be carefully defined for organic solvent denaturation, since the three stresses are all operating together on the proteins. With so many possible variables one is unlikely to ever reach the ideal procedure, i.e., maximum recovery of the required enzyme combined with maximum loss of other proteins. On the other hand, it is possible to make a large number of small-scale tests in a relatively short time so that a near-optimum set of conditons can be chosen.

A frequently used example of organic solvent precipitation, but this time with an immiscible solvent, is chloroform denaturation of hemoglobin (cf. Refs. 7 and, e.g., 36). Purification of erythrocyte enzymes would be complicated by the fact that well over 90% of the total protein in the red blood cell is hemoglobin. Shaking a hemolysate with ethanol-chloroform causes complete denaturation of the hemoglobin; a small percentage of chloroform mixes with the aqueous phase and has a strong, specific denaturing effect on the hemoglobin molecule. Thus the first stage in isolating an erythrocyte enzyme is often to carry out chloroform treatment, to remove the main contaminating protein, hemoglobin.

# Chapter 4
# Separation by Adsorption

Proteins adsorb to a variety of types of solid phases, usually in a selective manner. Consequently adsorption techniques, especially when adopted in column chromatography, have become widely used; they frequently result in purification steps which give the greatest increase in protein purity, that is, in the case of an enzyme isolation, the greatest increase in specific activity. Although column chromatography is the ideal way of getting optimum resolution, batchwise adsorption methods should not be forgotten, since they can be very rapid, and so are valuable techniques when speed is a priority (see also Chapter 7). The principal adsorbents for proteins are ion exchangers, calcium phosphate (as a gel or as a crystalline medium), and miscellaneous affinity adsorbents designed for particular types of enzymes. These are discussed in this chapter, following a general introduction to the theory of adsorption chromatography as applied to the isolation of a particular protein.

## 4.1   A Theory for Chromatography of Proteins

Chromatographic theory has been developed for low-molecular-weight compounds, and many of the ideas and principles are not always relevant to the types of interaction and behavior of proteins on adsorbents. Among the differences encountered with high-molecular-weight polymers are matrix exclusion effects, large ranges of interaction strengths with different parts of the adsorbent, and often high adsorption capacity (in g $cm^{-3}$, despite the deceptively low appearance when this is expressed in molar terms). Biochemists are accustomed to thinking of molecular interactions in terms of dissociation constants—which often can be related to $K_m$ values in enzyme kinetics. By using the same concept, of a dissociation constant for the interaction between adsorbent and protein, a simple theory of protein chromatography can be built up; this is particularly useful for describing affinity techniques.

Adsorbents for proteins have requirements different from those used for low-molecular-weight compounds. The adsorbent must have a very open structure so that the protein can penetrate the particles and reach the binding sites; if not, then only the few binding sites on the particle surfaces are available, and the capacity of the adsorbent is very low. Secondly, the nature of the adsorbent must not be harmful to proteins, nor should the backbone matrix material interact with the protein strongly—the adsorptive power is usually created by adding substituents to a supposedly inert backbone material. These two factors make the standard types of ion exchanger adsorbents for low-molecular-weight compounds useless for protein work—the highly cross-linked substituted polystyrene resins do not allow penetration by proteins, and they interact by hydrophobic as well as ionic forces on the exterior, which can make any adsorbed protein difficult to elute. In addition, the very high density of charged groups can make protein binding virtually irreversible. It has also been suggested that strongly acidic, e.g., sulfonic acid, or strongly basic substituents are harmful—however, this is a confusion of terminology; the charged form used in ion exchange work is the complementary weak base (e.g., sulfonate groups) or weak acid (e.g., quaternary ammonium groups).

The first successful materials used in column chromatography of proteins were the cellulose-based ion exchangers developed by Peterson and Sober (37). Many other adsorbents had been used for batch techniques, but few were suitable for column work because of poor flow characteristics. Cellulose powder suitably treated opens up to a porous structure which protein molecules can penetrate. Moreover, substitution of the cellulose with charged groups increases and stabilizes the porosity as a result of mutual repulsion between the charges. Such materials have proved ideal for protein ion exchange chromatography, and remain today the most widely used adsorbents. The other most important adsorbents for column work are also in the main based on biopolymers: dextrans or agaroses. These include the many true affinity adsorbents and the dye-substituent "pseudo-affinity" adsorbents, hydrophobic adsorbents, and alternative ion exchange materials. One exception is the crystalline form of calcium phosphate (hydroxyapatite), which can be used in a column; the more traditional calcium phosphate gel cannot. The chromatographic theory described below can be applied to all of these adsorbents, whether being used in a column or in a batch-wise method.

## The Dissociation Constant for Protein–Matrix Interaction

The interaction between a protein and an adsorbent is rarely one simple combination like that between an enzyme and its substrate. Many possible configurations of binding sites on the matrix present themselves to a protein, which itself may have an oligomeric quaternary structure and so offer more than one surface to bind. Figure 4.1 is a two-dimensional representation of how a protein might distribute itself in an adsorbent which has randomly distributed sites; it might interact with two, three, or more sites, giving a polydispersity of inter-action strengths. The + signs suggest an anion exchanger; however, the dia-gram can equally well represent most other types of adsorbent. Consequently an array of possible "dissociation constants" may be expected, not necessarily evenly distributed over the range. Is it possible to average out the spread of values and use a single figure which approximates the interaction of most mol-ecules? If not, the mathematical complexity of the situation and the necessity of defining a function to describe the distribution of constants would make this approach a near-impossible exercise. Fortunately, in most cases one can use a single "apparent" dissociation constant. Deviations from ideal behavior due to this approximation can be explained descriptively if not mathematically.

Experiments carried out with ion exchangers indicate that much of the bind-ing can be described by a single dissociation constant, making this approach a useful if only partially quantitative one (29,38). If we call the concentration of protein in free solution at equilibrium with the adsorbent, $p$, the concentration of free effective binding sites $m$ (effective binding sites means sites that can be occupied by protein without steric hindrance from adjacent bound protein, and whose affinity for protein is sufficiently high to have an observable effect), and the concentration of matrix-bound protein $q$, then the dissociation constant for the protein-matrix interaction $K_p$ is defined as

$$K_p = \frac{m \cdot p}{q} \tag{4.1}$$

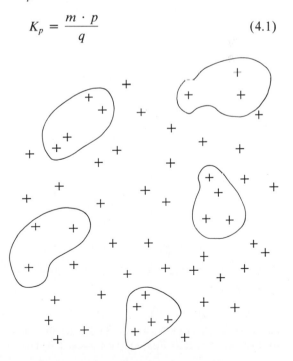

**Figure 4.1.** Two-dimension-al representation of the poly-dispersity of adsorbent sites in an ion exchanger. Protein molecules are shown inter-acting with three, four, and five positive charges (+) on the exchanger. The charged groups may be randomly distributed, or heterodis-perse due to inhomogeneities in the adsorbent.

In order to define these terms properly, the volume occupied by a given amount of matrix must be clarified—for a sensible definition we should assume that the adsorbent matrix is settled to a volume such as it would occupy when packed in a column. In these circumstances a typical modern adsorbent based on cellulose or agarose would have an effective liquid volume $(V_e)^*$ of 75–90% of the total $(V_t)$, the adsorbent plus bound water occupying the remainder. Thus the *amount* of protein in free solution around the adsorbent is somewhat less than its concentration times the *total* volume.

Having defined this dissociation constant, we can now use it in a way that usefully relates to adsorption chromatographic procedures. Adsorption represents a partitioning of solute (protein) between the two phases, liquid and solid. A partition coefficient can be defined as being the amount of solute adsorbed at any instant as a fraction of the total. Alternatively it could be defined as that fraction in solution, in which case it is identical with $R_f$, the mobility of the solute relative to the buffer front. The former will be used here and called $\alpha$ ($\alpha = 1 - r_f$). This partition coefficient $\alpha$ will be referred to frequently during the rest of this chapter. It should be noted that reference to a *low* $\alpha$ value implies zero to about 0.5, whereas a *high* $\alpha$ implies a value close to unity.

The relationship between $\alpha$ and $K_p$ can now be determined:

Let total effective concentration of adsorption sites (i.e., capacity for adsorbing the protein in question) be $m_t$. Let total protein concentration (free + bound) in the adsorbent be $p_t$. Then

$$m_t = m + q \qquad (4.2)$$

$$p_t = p + q \qquad (4.3)$$

$$\alpha = \frac{q}{p_t} \qquad (4.4)$$

Substituting $q = p_t\alpha$ and $m$, $p$ from Eqs. (4.2) and (4.3) into Eq. (4.1), we get

$$K_p = \frac{(m_t - p_t\alpha)(p_t - p_t\alpha)}{p_t\alpha} = \frac{(m_t - p_t\alpha)(1 - \alpha)}{\alpha}$$

So

$$p_t\alpha^2 - \alpha(m_t + p_t + K_p) + m_t = 0 \qquad (4.5)$$

Now $\alpha$ is the solution of this quadratic lying between 0 and 1. Note that in conditions where $m_t \gg p_t$, i.e., there is a large excess of binding sites compared with protein to bind, then

$$m \simeq m_t, \qquad q = p_t\alpha, \qquad \text{and} \quad p = p_t - q = p_t(1 - \alpha)$$

So

$$K_p = \frac{m \cdot p}{q} \simeq \frac{m_t \cdot (1 - \alpha)}{\alpha} \qquad (4.6)$$

---

$*V_e$ being the "elution volume" as defined in gel filtration theory, which is the volume accessible to macromolecules—the value depends on protein size.

Now $\alpha$ is a value that can be experimentally determined very simply; $p_t$ and $m_t$ are also easy to measure in specified conditions; thus the validity of these arguments can be experimentally tested, and values for $K_p$ determined. Typical concentrations of *effective* binding sites would vary between about 0.01 mM for affinity adsorbents to over 1 mM* for ion exchangers, and total protein concentrations would be in a similar range. Some solutions to Eq. (4.5) are presented in Table 4.1, using a variety of values for $K_p$. For a useful adsorption on column chromatography, $\alpha$ needs to be at least 0.8, from which it can be seen that the value of $K_p$ generally needs to be less than 0.1 mM. This has important ramifications when considering affinity adsorption techniques.

The conventional description of chromatographic processes makes use of the Langmuir isotherm, which expresses the amount bound ($q$) in terms of the concentration of free solute ($p$). Using the symbols defined above, this expression can be deduced from the dissociation constant $K_p$:

$$K_p = \frac{m \cdot p}{q} = \frac{(m_t - q) \cdot p}{q}$$

$$q(K_p + p) = m_t p, \qquad q = \frac{m_t p}{K_p + p}, \quad \text{or} \quad \frac{k_1 p}{1 + k_2 p} \qquad (4.7)$$

This is the Langmuir isotherm. Note that the concentration term $p$ is the *free* concentration at equilibrium, not the total. Thus the Langmuir isotherm does *not* represent a plot of amount adsorbed against amount *added,* as is sometimes thought. It is much more difficult to measure $p$ than $p_t$. Using

**Table 4.1.** Solutions for $\alpha$, from the Equation
$p_t\alpha^2 - \alpha(m_t + p_t + K_p) + m_t = 0$

| $K_P$ (mM) | $p_t$ (mM) | $m_t$ (mM) | $\alpha$ |
|---|---|---|---|
| 0.1 | 0.1 | 1.0 | 0.90 |
| 0.1 | 0.5 | 1.0 | 0.85 |
| 0.1 | 1.0 | 1.0 | 0.73 |
| 0.1 | 0.1 | 0.1 | 0.38 |
| 0.1 | 0.5 | 0.1 | 0.16 |
| 0.1 | 0.01 | 1.0 | 0.91 |
| 0.01 | 0.01 | 1.0 | 0.99 |
| 0.01 | 0.5 | 1.0 | 0.98 |
| 1.0 | 0.01 | 1.0 | 0.50 |
| 0.1 | 0.1 | 10 | 0.99 |
| 1.0 | 0.1 | 10 | 0.99 |
| 0.1 | 0.1 | 0.2 | 0.59 |
| 0.1 | 0.1 | 0.5 | 0.81 |
| 0.1 | 0.1 | 2 | 0.95 |
| 0.1 | 0.1 | 5 | 0.98 |

*Highly dependent on protein size; see Table 4.3.

some convenient values, the Langmuir isotherm expression is compared with the amount adsorbed per amount added and with $\alpha$ in Figure 4.2.

Column chromatographic techniques typically involve applying a relatively large sample—often in a volume much greater than the volume of the column—in a buffer chosen so that the protein of interest adsorbs. In the starting conditions there are three possible fates for a particular protein. Either it adsorbs totally and immovably ($\alpha = 1$), it does not adsorb at all ($\alpha = 0$), or it adsorbs partially ($0 < \alpha < 1$) and moves down the column, emerging when $1/(1 - \alpha)$ column volumes of liquid following the start of application have passed through. Thus if the applied sample is 5 times the column volume, and the column is to be washed with a further 5 vol, for this protein to remain in the column, $\alpha$ would need to be greater than 0.9. On the other hand, if $\alpha = 0.4$ and only a small sample is applied to the column, a very convenient purification method might be achieved (Figure 4.3). As will be shown later, $\alpha$ changes very rapidly with minor variations in conditions, and it is unlikely that another major protein would have the same intermediate value. If the enzyme of interest has an $\alpha$ value of 0 or only a little greater than zero, then it passes through the column, along with all other proteins of $\alpha = 0$. This may be a good purification step, especially if the majority of proteins are left behind adsorbed to the column. No true chromatographic principles are involved in this case, other than enabling proteins with high but *not unity* values to be totally adsorbed to the column—this would not be possible with batch techniques (cf, section 4.9). The enzyme in the nonadsorbed fraction would be obtained rapidly, since the column does not need to be developed with an elution scheme; it should be obtained in 100% yield. It would also probably be in a low ionic strength buffer which may be suitable for an immediate application and adsorption to another type of chromatographic column. Thus, although the full

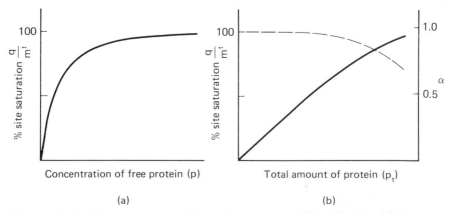

Concentration of free protein (p)                Total amount of protein ($p_t$)

(a)                                                              (b)

**Figure 4.2.** (a) The Langmuir isotherm plot, expressing the percentage of site saturation (amount adsorbed) against *free* solute concentration. (b) Percentage site saturation (solid line) and the partition coefficient $\alpha$ (dashed line) against the *total* solute added.

**Figure 4.3.** Idealized behavior of a mixture of proteins with $\alpha = 0$ and one of $\alpha = 0.4$. Here $\alpha = 0$ proteins emerge at 1 column volume of eluant; $\alpha = 0.4$ protein emerges at $1/(1 - \alpha) = 1.7$ column volumes (shaded peak). No change of buffer conditions is involved.

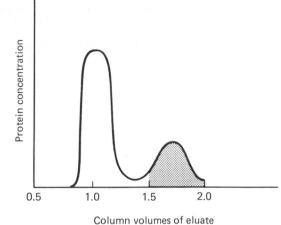

Column volumes of eluate

benefits of chromatography can only be obtained by first adsorbing all the enzyme ($\alpha > 0.8$), then eluting with some scheme involving changing buffers, at least partial purification may be achieved in other circumstances.

Even if all adsorption sites were equivalent, the $\alpha$ value would depend on protein concentration [see Eq. (4.5), Figure 4.2b], with higher protein concentrations leading to a smaller *proportion* (though larger amount) adsorbed. The fact that there is in truth a whole range of $K_p$ values further aggravates this effect. Stronger binding sites (low $K_p$) are occupied first; higher protein concentrations need more, weaker binding sites, leading to a higher average $K_p$ and a lower $\alpha$. Some quantitative values for rabbit muscle aldolase adsorbing to CM-cellulose have been presented elsewhere (38). These results are illustrated in Figure 4.4, in which the variation of $\alpha$ with the total amount of protein present is plotted for four different pH values. If we consider a band of aldolase on a column at pH 7.5, the total protein concentration in the middle of the band may approach $m_t$, with an $\alpha$ value of about 0.6–0.7. Any protein that moves out ahead of the main band, whether due to diffusion or to uneven flow in the column, becomes diluted and the $\alpha$ value is higher. At $p_t/m_t = 0.1$, $\alpha = 0.9$ in this case. Consequently it mostly adsorbs, and so is retarded relative to the main band behind it. As a result there is a self-sharpening effect on the leading edge of the protein band. However, on the trailing edge, decreasing protein concentration causes an ever-increasing strength of binding. Now $\alpha$ approaches 0.9, so that the rear part of the protein band moves with only 10% of the rate of the buffer flow, compared with 30–40% for the main band. As a result, with constant buffer conditions the enzyme would elute from the column with a sharp, concentrated leading edge, but a very long tail (Figure 4.5). This is an occurrence frequently observed with ion exchangers, but it can be overcome using a continuously changing gradient of buffer conditions (cf. section 4.3).

It can be calculated from Figure 4.4 that the measured average $K_p$ values

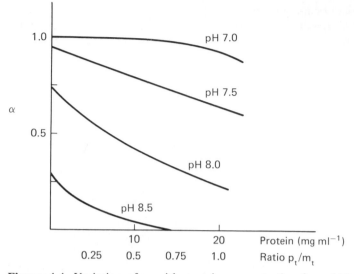

**Figure 4.4.** Variation of $\alpha$ with protein concentration for rabbit muscle aldolase adsorbing to CM-cellulose, at several pH values. [Experimental results from Scopes (38).]

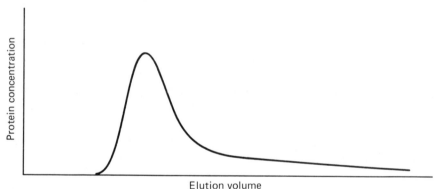

**Figure 4.5.** Shape of elution peak in constant buffer conditions expected for rabbit aldolase at pH 7.5 (from Figure 4.4). The long "tail" is caused by increasing value of $\alpha$ as the protein becomes more dilute.

change by a factor of about 5 from very low site occupancy to 50% occupancy. The value for the total available sites, $m_t$, may not be the same in all conditions. The total effective capacity of an ion exchanger may well be greater when the binding is very tight, and less when the binding is weaker, despite the fact that the adsorbent itself presents the same pattern and distribution of possible adsorbing sites. A protein with 50 net charges may be able to bind on a "weak" site, whereas when it has only 30 net charges it may not have any significant interaction.

Another significant complication here is that there is a proportion of adsorbing sites which take quite a long time to be occupied. Indeed, the time dependence is such that total saturation may take as long as an hour to be reached. In these cases equilibrium treatment is clearly not appropriate. These slow, poorly accessible sites may become occupied during the long process of loading a column. If it takes a long time for the protein to go on, it usually follows that it must also take a long time for it to come off. Thus a rapid elution scheme will leave behind some small proportion of enzyme, but it elutes later—a further aggravation of the tailing effect already described in terms of equilibrium consideration of partition coefficients and variable dissociation constants. Some protein may never come off, especially if a change in buffer conditions leads to a shrinkage of the adsorbent, finally trapping the protein in a site that was already accessible only with difficulty.

The final complication that upsets theoretical predictions is the occurrence of inhomogeneity of adsorbents; this is clearly a principal factor in causing the range of $K_p$ values, but it also means that the concentration of binding sites, $m_t$ (Eq. 4.2), is only an average value, as it is calculated from the concentration of bound protein. In some adsorbents it may be that large volumes of matrix are so poorly substituted that there are few effective binding sites present, but where substitution does occur it is so extensive that virtually all the protein that binds does so in this relatively small fraction of the total volumne of adsorbent. Consequently the true, effective value of $m_t$ to put into Eq. (4.5) may be much higher than the average value calculated from protein binding per cubic centimeter. This would result in higher $\alpha$ values than if the sites were distributed evenly through the matrix. Further discussion of this point occurs later under affinity adsorption (cf. section 4.5).

In summary, the description of adsorption effects by using a dissociation constant for the protein-matrix adsorption can be useful, bearing in mind that in most cases a unique value for the dissociation constant does not exist, and the average value will vary depending on the percentage of adsorption sites occupied. Similarly the partition coefficient, which is the experimentally important parameter, can decrease both directly due to approach to saturation, and indirectly because increasing site occupancy results in a lower average affinity, giving a lower $\alpha$ value.

# 4.2  Ion Exchangers—Principles, Properties, and Uses

Although the previous section was a general one describing the principles of chromatography, ion exchangers were mentioned frequently, and much of the theory has been tested using ion exchange adsorbents. Proteins bind to ion exchangers by electrostatic forces between the proteins' surface charges (mainly) and the dense clusters of charged groups on the exchangers. The substitution level of a typical diethylaminoethyl- (DEAE-) cellulose or carboxy-methyl- (CM-) cellulose may be as much as 0.5 mmol cm$^{-3}$ (packed, swollen adsorbent), that is, 0.5 M of charged groups. The charges are of course balanced by counterions such as metal ions, chloride ions, and sometimes buffer ions. A protein must displace the counterions and become attached; generally the net charge on the protein will be the same sign as the counterions displaced—hence "ion exchange." The protein molecules in solution also are neutralized by counterions; the overall effect in a given region of the adsorbent must be electrically neutral—this is illustrated in Figure 4.6. In this diagram it has been assumed that a protein with a net negative charge is preequilibrated in Tris-chloride buffer: the counterions associated with the protein are thus HTris$^{+}$. The adsorbent (DEAE-cellulose) also equilibrated with the buffer has Cl$^{-}$ counterions.

The protein then displaces the chloride ions, occupies a site in the adsorbent, and Tris Cl is discharged. It may be noted that this Tris Cl is the (acid) salt of Tris, and if the displacement is rapid, it could lead to a lowering of the effective pH in the column ahead of the band of adsorbed protein.

In addition, the pulse of Tris Cl represents an increase in the ionic strength of the buffer flowing through the column. Typically for every 1 mg ml$^{-1}$ protein that is adsorbed, it is replaced by approximately 1 mM of extra buffer salts.

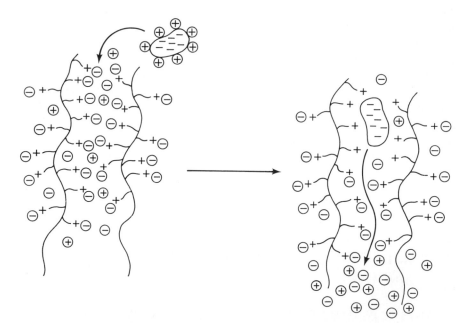

**Figure 4.6.** Illustration of the "ion exchange" occurring when a negatively charged protein adsorbs to an anion exchanger. Seven positively charged ions (e.g., HTris$^{+}$) associated with the protein molecule are displaced, together with seven negative ions (Cl$^{-}$) from the exchanger.

The pH of nonadsorbed fractions can be shown to be lower than the preceding buffer wash (or higher, when referring to a cation exchanger) if the protein concentration of the applied sample is high. To avoid these pH and ionic strength changes, which can result in less adsorption than expected, the applied protein concentration should not be too high—especially if a large proportion of it is expected to adsorb to the column. This also implies that there should be adequate buffering—in the absence of any buffer, quite extreme pH changes could occur and cause denaturation. Generally 10 mM of a buffer within 0.3 unit of its $pK_a$ is the minimum desirable level (cf. Table 4.6 and section 6.1) and protein *being adsorbed* should not be more than about 5 mg ml$^{-1}$. (Proteins that are not being adsorbed have less influence; but cf. section 4.3.)

Two forms of interaction have been suggested for the adsorption of proteins to ion exchangers. One is termed "irreversible," the other "reversible." As has been pointed out by Peterson (39), the term *irreversible* is unfortunate in implying that the protein, once adsorbed, will not come off again: this would not be very useful. A smooth change of buffer conditions to cause the transition from so-called irreversible to reversible adsorption must in fact be a gradual process. The probability that a protein molecule will dissociate in one second may be 99% for "reversible" adsorption and 0.001% for "irreversible" (it can never genuinely be zero), but where the dividing line comes is really a matter of opinion. These probabilities could be calculated from adsorption/pH curves such as illustrated in Figure 4.4, but ultimately one ends up with just another way of expressing the partitioning between protein and ion exchanger. A recent series of experiments, similar to those illustrated in Figure 4.4 demonstrate the transition from adsorbed (partition coefficient close to unity) to nonadsorbed (partition coefficient zero) as the pH of the buffer changes (38). Calculations from these curves using the equations in section 4.1 have given some quantitative estimates of the magnitude of dissociation constants (assuming quasi-equilibrium conditions) between protein and adsorbent. In these examples the adsorbent studied was CM-cellulose. Not only was it possible to calculate the values for $\alpha$ and $K_p$, but also from the rate of change of these with pH a value for the energy of interaction per charge on the protein was obtained. (In practice this is the rate of increase in the energy of interaction for each additional positive charge, since assymetries of charge distribution and uncertainties about exact isoelectric points make it difficult to ascertain exactly how many charges on the protein molecule are really contributing to the interaction with the adsorbent.) The derivation of these parameters is shown below and in Figure 4.7.

From a pH/titration curve and a knowledge of the protein's isoelectric point, the number of charges on the molecule at each pH can be calculated. A plot of log $K_p$ against the number of charges on the protein molecule, $z$, gives a slope of $s$ (see Figure 4.7), where $s = (d \log K_p)/dz$.

The energy of interaction per mole is given by $\Delta G^0 = -2.3RT \log K_p$.

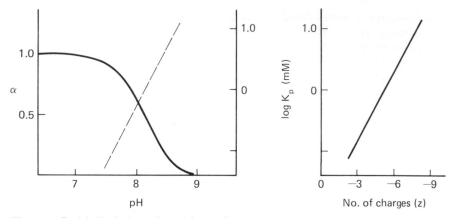

**Figure 4.7.** (a) Variation of $\alpha$ with pH for yeast phosphoglycerate kinase on CM-cellulose, $I = 0.01$ (solid line). Values for $K_p$ have been calculated using Eq. (4.6) (dashed line). (b) log $K_p$ plotted against number of charges on the enzyme molecule, calculated from a titration curve and taking the isoelectric point as 7.1. The value of $s$ (see text) is 0.4.

Therefore, the rate of change of energy of interaction per mole per charge $z$ is

$$\frac{dG^0}{dz} = -2.3RT\frac{d\log K_p}{dz} = -2.3RTs$$

Values for $\dfrac{dG^0}{dz}$ for various proteins are listed in Table 4.2. It can be seen

seen that the values show little definite trend; they are mostly in the range $1-2$ kJ mol$^{-1}$. However, the smallest protein, myokinase (MW 21,500) had the largest value, indicating that the extra positive charges that the protein acquired as it titrated through the range pH 7.5–6.0 were interacting more strongly with the ion exchanger than was the case with larger protein molecules. Either this is due to the fact that the smaller molecule is able to pack more tightly into the exchanger (confirmed by capacity experiments, see below) or that in a small molecule, each charge is inevitably closer on aver-

**Table 4.2.** Change in the Energies of Interaction of Enzymes with CM-Cellulose for Each Positive Charge on the Enzyme

| Enzyme | $dG°/dz$ (kJ mol$^{-1}$) |
| --- | --- |
| Yeast phosphoglycerate kinase | 2.0 |
| Rabbit muscle AMP kinase | 4.2 |
| Rabbit muscle creatine kinase | 1.5 |
| Rabbit muscle aldolase | 0.9 |
| Rabbit muscle pyruvate kinase | 1.0 |
| Yeast pyruvate kinase | 1.7 |
| Rabbit muscle lactate dehydrogenase | 0.7 |
| Beef heart mitochondrial aspartate amino transferase | 0.9 |

age to the protein molecule's surface, and therefore forms a stronger electrostatic attraction with the ionized carboxymethyl groups of the adsorbent. To estimate the number of these ionized groups that are interacting with each charge in the protein, one can use appropriate values for likely distances between the average residue in a protein and the ionized carboxymethyl groups, use the dielectric constant $D$ for water, and use a value of 2 kJ mol$^{-1}$ per charge in Eq. (4.8)

$$E = \frac{-z_1 z_2}{Dr} \qquad (4.8)$$

where $z_1$, $z_2$ = charges on attracting components—$z_1$ for protein; $z_2$ for adsorbent; $r$ = distance between them; taken as 5 nm

$$E(\text{per charge } z_1) = \frac{-z_2}{Dr} = 2 \text{ kJ mol}^{-1}$$

This gives $z_2$ = approximately $-5$, i.e., on average about 5 carboxymethyl groups are exerting their influence on each positive charge. These calculations are based on extremely broad assumptions, but at least the value for $z_2$ is reasonable, when considering the probable three-dimensional density of charge clusters of carboxymethyl groups in relation to the sizes of proteins.

## Adsorptive Capacities of Ion Exchangers

The amount of protein that can be bound per unit volume of packed ion exchanger can be very high indeed. However, for classical style exchangers such as cellulose-based materials it depends greatly on the size of the protein molecule, with the smaller molecules adsorbing to a greater extent. Expressed in molar terms the differential is still greater. Some typical values are given in Table 4.3. Molecules with molecular weights greater than $10^6$ are likely to be excluded from most cellulose-based ion exchangers, the exclusion effect being essentially the same as is occurring in gel filtration chromatography (cf. section 5.1). Larger molecules can only bind to the surface of the ion exchanger particles, and so the capacity for these is very low. The total capacity relates to the amount of the internal volume of the particles that is accessible to the protein, and as with gel filtration, this is simply a hole-size problem. As shown in Figure 4.8, the capacity, in mg cm$^{-3}$, increases with decreasing molecular weight in such a way that capacity against log MW is nearly linear. It is interesting to reflect that an adsorbent already "saturated" with a large protein molecule is still capable of taking up small molecules, provided that the "pores" are not sterically blocked up by the large molecules. In an experiment to confirm this, using the conditions described in Figure 4.8, with CM-cellulose at pH 6.0, 1 cm$^3$ of CM-cellulose was treated with an excess of rabbit muscle pyruvate kinase (MW 228,000); 13 mg was adsorbed. The CM-cellulose was washed with buffer, then 50 mg lysozyme (MW 14,500) was added. All the lysozyme adsorbed to the CM-cellulose, without displacing the pyruvate

**Table 4.3.** Capacity of CM-Cellulose (Whatman CM 52) for Various Proteins at $\alpha$ = 1.0 (pH 6.0, $I$ = 0.01)[a]

| Protein | MW | Capacity (mg cm$^{-3}$) | Capacity $m_t$ (mM) |
|---|---|---|---|
| Hen egg white lysozyme | 14,300 | 130 | 9.0 |
| AMP kinase | 21,400 | 100 | 4.6 |
| Yeast and rabbit muscle phosphoglycerate kinase | 45,000 | 70 | 1.55 |
| Phosphoglycerate mutase | 60,000 | 40 | 0.67 |
| Creatine kinase | 82,000 | 35 | 0.43 |
| Enolase | 88,000 | 48 | 0.55 |
| Lactate dehydrogenase | 140,000 | 21 | 0.15 |
| Glyceraldehyde phosphate dehydrogenase | 145,000 | 26 | 0.18 |
| Aldolase | 160,000 | 22 | 0.14 |
| Pyruvate kinase | 228,000 | 12.5 | 0.055 |

[a]See Figure 4.8 for a plot of these values.
*Note:* All enzymes are from rabbit muscle unless otherwise indicated.
*Source:* Scopes (38).

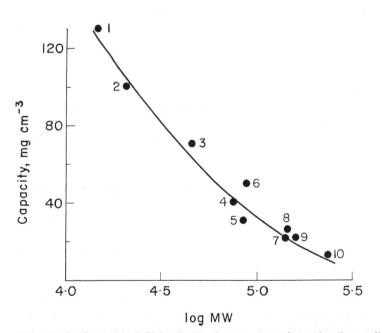

**Figure 4.8.** Capacity of CM-cellulose for a variety of proteins (in conditions where $\alpha$ = 1.0). Proteins used: (1) hen egg white lysozyme; (2) AMP kinase; (3) phosphoglycerate kinase (both rabbit muscle and yeast preparations gave the same capacity); (4) phosphoglycerate mutase; (5) creatine kinase (at 0°C); (6) enolase; (7) lactate dehydrogenase; (8) glyceraldehyde phosphate dehydrogenase; (9) aldolase; (10) pyruvate kinase. [From Scopes (38).]

kinase. This illustrates that overloading of columns may result in the emergence of the larger molecular size protein first; there may be sufficient capacity to adsorb all the smaller molecules—depending on the relative proportions of large and small molecules in the applied sample. Newer ion exchangers have higher exclusion limits; LKB's Trisacryl is claimed to be able to take up molecules up to $10^7$ in molecular weight (40).

## Types of Ion Exchangers

A large variety of types of ion exchangers have been developed over the years, but many of them have found limited use or have been superceded by superior materials. The beginner need have only two types, a cation exchanger (carboxymethyl) and an anion exchanger (DEAE). Although phosphocellulose has found wide application, there is much evidence that this is a combined ion exchanger/affinity adsorbent, with special affinity for enzymes with phosphate ester substrates—thus phosphocellulose fits into a category of its own. Carboxymethyl and diethylaminoethyl groups have been attached to cellulose, agarose, or dextran to provide suitable materials for protein ion exchange chromatography. Early materials were coarse cellulose particles which had such desirable features as rapid flocculation and good flowrate in columns, but because of their microstructure there were relatively few binding sites for proteins—especially large proteins—so their capacity was limited. An improved "microcrystalline" cellulose material was developed by Whatman, and their two exchangers CM 32 and DE 32 (dry powders), or CM 52 and DE 52 (damp, preswollen versions), have found widespread use, and are still competitive, at least in price, with the more sophisticated materials developed since.

With these substituents attached to spherical dextran beads (Sephadex—Pharmacia), high-capacity adsorbents were available with reliable, reproducible properties. However, these were not very successful because of their propensity to shrink with changes of buffer conditions. In water the soft dextran beads swell enormously—the suspension is almost transparent—due to repulsive charge forces within the substituted beads. These repulsive forces weaken as salt concentration rises, and the beads collapse—with unfortunate effects in a column. Not only does the shrinkage of the column cause technical problems of increased dead space and irregular gradients, but the shrinkage of the beads can trap proteins—especially the larger ones—so that they are never eluted. Thus the dextran ion exchangers should only be used when either the application of sample can be made at a relatively high ionic strength ($>0.1$) or when ionic strength is to remain constant and elution carried out by pH changes and/or affinity elution (cf. section 4.4). More successful materials from Pharmacia (and from other suppliers) are the substituted cross-linked agarose gels, such as DEAE-Sepharose. The agarose gel beads have a large exclusion size, so they are suitable for proteins up to $10^6$ daltons, and being cross-linked they are rigid, showing little change in volume under any conditions. Pharmacia has also introduced spherical cellulose beads; to date only

DEAE-substituted Sephacel is available. Some of the more popular ion exchangers are listed in Table 4.4.

So far no mention has been made of other substituents, mainly because one cation exchanger and one anion exchanger should be able to do all that is necessary in the vast majority of cases; too many choices of materials can lead to a lot of wasted time trying each out. However, alternative materials and their properties will be described now. At extremes of pH, both CM and DEAE substituents become uncharged. If it is appropriate to operate an anion exchanger above pH 9, or a cation exchanger below pH 4 (and relatively few enzymes are stable in these conditions), other substituents are needed. Sulfoethyl or sulfopropyl groups which remain ionized down to pH 2 have been attached to celluloses or dextran beads. For anion exchangers, quaternary ammonium groups remain ionized even at pH 12. Titration curves for substituted Sephadexes in 1 M KCl are presented in Figure 4.9. Although QAE-Sephadex has a small inflexion around pH 7, it is essentially fully positively charged at all pH values below 12; similarly SP-Sephadex is fully negatively charged at all pH values above 2. CM-Sephadex (and other CM-substituted materials) titrates around pH 4.5, and can be considered fully negatively charged over pH 5.5. DEAE-substituted materials (this applies to celluloses and agarose as well as dextrans) are more complex, and can be regarded as fully positively charged only below pH 5. Whereas the diethylaminoethyl group has a $pK_a$ of 9 to 9.5, substantial titration of DEAE-substituted material is noted around pH 6–8 (Figure 4.10). This is due to closely adjacent substituents affecting each others' $pK_a$ by electrostatic repulsion of protons. Thus DEAE-cellulose, the most com-

**Table 4.4.** Some Commercially Available Ion Exchange Products

|  | Anion exchangers | Cation exchangers |
|---|---|---|
| Cellulose-based | DEAE[b,c] | CM[b,c] |
|  | TEAE[b] | phospho[b,c] |
|  | QAE[b] |  |
| Sephacel (spherical cellulose beads) | DEAE[a] |  |
| Sephadex (dextran beads) | DEAE[a] | CM[a] |
|  | QAE[a] | Sulfopropyl[a] |
| Agarose-based | DEAE[a,b] | CM[a,b] |
|  | PEI (Polybuffer exchanger)[a] |  |
| Synthetic-polymer-based (Trisacryl) | DEAE[d] | CM[d] |

[a]Pharmacia.
[b]Bio-Rad Laboratories.
[c]Whatman.
[d]LKB.
*Key:* DEAE—diethylaminoethyl; TEAE—triethylaminoethyl: QAE—diethyl-(2-hydroxypropyl) aminoethyl; PEI—polyethylene imino; CM—carboxymethyl.

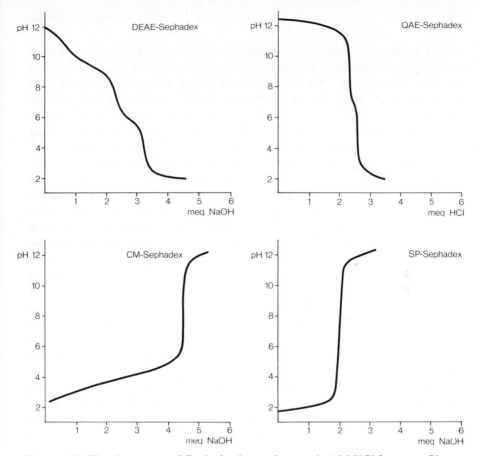

**Figure 4.9.** Titration curves of Sephadex ion exchangers in 1 M KCl [courtesy Pharmacia Fine Chemicals (41).]

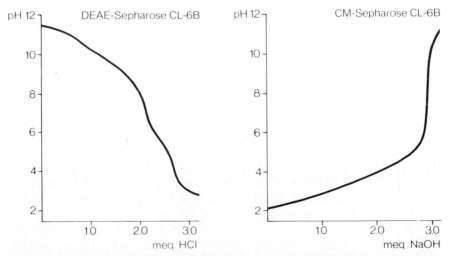

**Figure 4.10.** Titration curves for DEAE-Sepharose and CM-Sepharose in 1 M KCl [courtesy Pharmacia Fine Chemicals (41).]

monly used ion exchanger, has a complex structure and titration curve, and in normal operating buffers around pH 6–9 some groups on the adsorbent are only partially ionized. This does not detract from its usefulness; since most protein elution schemes are developed empirically, the underlying properties may not be of much concern. However, it is worth noting that at lower pH values, whereas fewer proteins will be sufficiently negatively charged to adsorb to DEAE substituents, the *capacity* of the adsorbent for those that do bind will be greater. Thus the adsorbent is much less likely to become overloaded if a lower pH buffer can be chosen.

QAE-Sephadex and SP-Sephadex suffer the problems of swelling and contraction referred to above. A cellulose substituted with a quaternary ammonium group triethylaminoethyl, TEAE-cellulose, has been available for some time; however, its titration does not differ much from that of DEAE-cellulose, and it must be concluded that few of the groups are in fact quaternary ammonium. Other substituents such as ECTEOLA and aminoethyl have largely gone out of use. Two ion exchangers (neither of which is presently available) would be sufficient and ideal: a quaternary ammonium-substitued cross-linked agarose (or spherical cellulose bead or organic polymer-based material) for anion exchange and sulfopropyl substituted agarose for cation exchange. But there are cases where an adsorbent has been chosen because it has a *low* capacity for proteins—the protein desired binds very tightly, to the exclusion of others. On a highly substituted ion exchanger the tightness of binding might be so great that the protein cannot be removed. For these cases a standardized material with a stated lower substitution could be available. The reason why a lower degree of substitution not only lowers the capacity but also the tightness of binding is because an electrostatic interaction between a protein and an adsorbent with fewer substituents per unit volume is not as strong.

## pH and Donnan Effects

The pH in the microenvironment of an ion exchanger is not the same as that of the applied or eluting buffer because Donnan effects can repel or attract protons within the adsorbent matrix. In general, the pH in the matrix is usually about 1 unit higher than the surrounding buffer in anion exchangers and 1 unit lower in cation exchangers (Figure 4.11). The lower the ionic strength of the buffer, the larger this difference is. This has important ramifications when considering the stability of enzymes as a function of pH; an enzyme which is perfectly stable at pH 5.5 may denature rapidly in solution at pH 4.5. If such an enzyme is adsorbed on a cation exchanger at pH 5.5, it experiences a microenvironmental pH lower than 5.5, and is likely to denature on the adsorbent. Because it leads to a shift toward extreme pH values, the Donnan effect severely limits the operational pH range of ion exchangers, especially in the mildly acid range that is needed on cation exchangers. In general, enzymes tend to be more stable in the mildly alkaline range, e.g., pH 8–10, than mildly acid, pH 6–4, so loss by denaturation is less of a problem using anion exchangers.

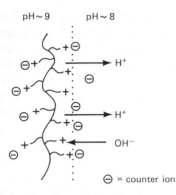
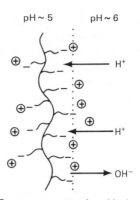

Protons are expelled, and hydroxyl ions attracted to the immediate environment of anion exchanger

Protons are attracted, and hydroxyl ions expelled from immediate environment of cation exchanger

**Figure 4.11.** The Donnan effect on pH in the microenvironment of ion exchangers.

## Elution of Adsorbed Protein

In theory, two general methods for eluting proteins are available (plus affinity elution; cf. section 4.4). These are (a) to change the buffer pH to a value where binding is weakened—lower pH for an anion exchanger, or higher pH for a cation exchanger; and (b) to increase the ionic strength, thereby weakening the electrostatic interactions between protein and adsorbent. In practice, method (a) is not generally very successful. This is because, unless there is a very high buffering capacity, sudden large pH changes as proteins become eluted result in little separation of individual components. At low ionic strength, buffering capacity must be low, and the attempt to change the pH by applying a pH gradient is frustrated by the buffering power of the proteins adsorbed to the column and, in the case of DEAE-adsorbents, the buffering of the adsorbent groups themselves. A typical result is shown in Figure 4.12. Only if the proteins required are adsorbed very strongly in the first place can a pH gradient be used successfully. In this case strong buffering at ionic strength 0.1 or greater can be employed. A scheme for obtaining a strongly buffering pH gradient at constant ionic strength has been previously suggested (29).

Recently a new development in pH elution from ion exchangers has made this approach considerably more successful. Sluyterman et al. (42–45) have developed a technique called "Chromatofocusing," in which the pH gradient is formed using ampholyte-type buffers, of high buffering power but low effective ionic strength, and an exchanger (polyethyleneimine-agarose) which, though similar to DEAE-adsorbents, has a continuous titration of its groups through a large pH range. Pharmacia is marketing this system using their Polybuffer and Polybuffer exchanger (46). The pH gradient is not formed before running on to the column, but is generated within the column by continuous addition of an acid form of the ampholyte, Polybuffer. As a result, a very steady pH gradient emerges from the column, at low ionic strength, and

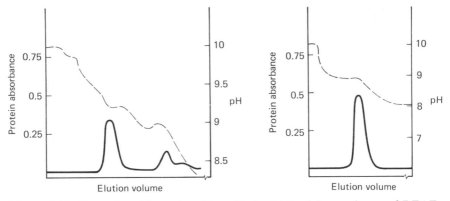

**Figure 4.12.** Examples of irregular pH gradients obtained from columns of DEAE-Sephadex. After starting at pH 10.1, buffers at pH 8.0 (a) or pH 7.5 (b) were applied to the columns. Elution pH values fell irregularly and dropped sharply as the protein (solid line) was eluted. [From Sluyterman and Wijdenes (44).]

proteins emerge at or somewhat above their isoelectric points. The resolving power is very high, in many instances higher than the more conventional salt gradient to be described below. It is claimed that each component can be totally eluted within a pH range of 0.05 units; this is probably an optimum which the average protein does not reach. The main examples given are proteins which have a good titration around their isoelectric point, a necessary feature for "isoelectric" resolution (Figure 4.13). Undoubtedly this method of elution, even from conventional ion exchangers, will be used extensively in the future, although considerations of cost effectiveness will have to be taken into account for large-scale work.

Salt gradients are the commonest means of eluting proteins from ion exchangers. Whereas pH changes can be "held up" by buffering action of proteins and column substituents, salts pass through the column unhindered, unless the salt used involves a change of counterion. Even in that case, the salt concentration remains the same, though its composition alters. Usually either potassium or sodium chloride is used to generate the gradient, and a linear gradient is formed by a simple mixing device (see Figure 4.19). The action of the salts can be considered in one of two ways. The salt can directly displace the protein; the ions (e.g., chloride ions on DEAE-adsorbents) occupy the positively charged sites and block reattachment by protein. Alternatively, the system can be regarded as an equilibrium in which even strongly bound proteins spend some time not adsorbed; the presence of the salt ions between the unattached protein and the adsorbent greatly weakens the attraction between the two. In either case, the desorbed protein is replaced by counterions, and so "ion exchange" is a true description.

As salt concentration increases, the value of $\alpha$ decreases. On a column this means that proteins begin to move down, and as they will be moving at a slower rate than the salt passing by, each part of the protein band is continuously

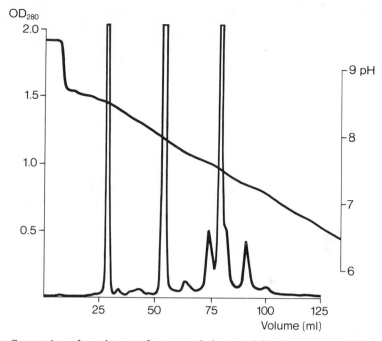

**Figure 4.13.** Separation of a mixture of sperm whale myoglobin, horse myoglobin, and carboxyhemoglobin (2 mg each) on Chromatofocusing system [courtesy Pharmacia Fine Chemicals (46).]

experiencing an increasing salt concentration. This causes a sharpening of the bands. The leading edge is sharpened as a result of effects described earlier, as well as by the fact that there is a gradient of conditions in the column making the potential $\alpha$ value still higher ahead of the band (Figure 4.14). The trailing edge, which would in nongradient conditions lag behind because of an increase in $\alpha$, is speeded up by the higher salt level. The actual $\alpha$ value may be no

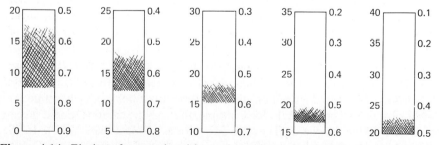

**Figure 4.14.** Elution of a protein with a salt gradient. On the left of each column diagram the salt concentration in the column is given (mM); on the right is the corresponding $\alpha$ value (for simplicity, variation in $\alpha$ with protein concentration has been ignored). The protein band moves faster as the salt increases, and emerges in this case at $\alpha = 0.5$.

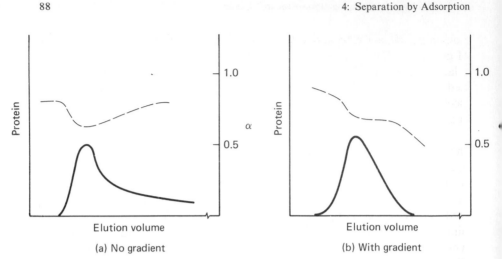

**Figure 4.15.** Elution of a protein (solid line) with a salt gradient. The variation of $\alpha$ with protein concentration means that without a gradient (a) considerable "tailing" of the peak occurs. With a gradient (b) the "tail" is accelerated and eluted quickly.

higher, or even lower immediately behind the main part of the band. The sharpening effect of a gradient is illustrated in Figures 4.14 and 4.15. The usefulness of stepwise increases in salt concentration to elute proteins is described in the next section.

## 4.3 Ion Exchange Chromatography—Practical Aspects

The isoelectric point of a protein depends on the proportions of ionizable amino acid residues in its structure. Positive charges are provided by arginines, lysines (the majority positive below pH 9.5), and histidines (positive below pH 7). Any free N-terminal amines will also bear a positive charge below pH 8. Unless one is operating at an unusually high pH, e.g., above 8.5, all arginines and lysines can be considered positively charged. Negative charges are principally on aspartate and glutamate residues, and above pH 6 virtually all these residues are ionized as are C-terminal carboxyls. At high pH values ($>8$) cysteines may become ionized also. A few residues, usually associated with the active site of an enzyme, may prove exceptions to these generalizations, with $pK_a$ value several units away from normal. For example, the enzyme creatine kinase has an ionized cysteine residue at its active site with a $pK_a$ less than 7 (47). In the normal pH range of enzyme stability on ion exchange columns, namely 5.5–9, the variation in net charge is mainly due to histidine residues, since these have $pK_a$ values in the range 6–7. Consequently a protein with few histidines will behave very similarly on ion exchangers over quite a wide pH range—the pH would not be critical. Conversely, proteins with high histidine content vary considerably in net charge through this pH range, and the precise pH for adsorption or elution would be more critical. Above pH 9 and below pH 5.5, titration of groups on proteins is extensive, and a very small variation in pH can make a large difference to the adsorption behavior ($K_p$ and $\alpha$ values). Often the limited range of enzyme stability restricts the pH that can be used, and if the isoelectric point is outside the stability range, only one class of ion exchanger can be used, as indicated in Table 4.5.

However, this assumes that all proteins behave ideally and do not adsorb on the "wrong side" of their isoelectric point, and that useful values of $\alpha$ ($>0.9$) are not reached until 1 or 1.5 pH units of the "right side." There are several documented cases where this simple concept proves to be incorrect—proteins adsorbing to cation exchangers even 1 pH unit above their isoelectric point (48), or vice versa for anion exchangers. There are "sticky" proteins which will bind to both types of exchanges in the same buffer, and "slippery" proteins which are difficult to get to bind to any adsorbent within a stable pH range. These effects are due partly to nonelectrostatic interactions, e.g., hydrophobic and van der Waals forces and partly to uneven distribution of charges over the protein surfaces. A cluster of like-charged groups, particularly at an active site which binds a multiple-charged substrate, can cause adsorption despite an overall average charge on the protein of the other sign (29). This effect is described in further detail in section 4.4.

## Trials to Determine Ion Exchange Behavior

In practice, the conditions for adsorption will be found empirically. Although a preliminary experiment of analytical isoelectric focusing combined with specific enzyme staining can reveal the isoelectric point, a simple experiment using ion exchangers will suffice equally well. A small sample of a preparation containing the enzyme can be passed through a gel filtration column to change the buffer. One ml can be processed through a 10-ml column in about 10 min (see Table 1.2). For a trial DEAE-cellulose adsorption, the buffer should be 20 mM Tris pH 8.0 (see below for buffers). A small column of about 2 cm³ should be sufficient, preequilibrated in the Tris buffer. If the enzyme does not adsorb, then (a) probably the process has resulted in a good deal of purification, since many other proteins are likely to have adsorbed to the DEAE-cellulose, and (b) the enzyme will probably adsorb to CM-cellulose. A second trial, using 20

**Table 4.5.** Choice of Ion Exchanger for Purification of a Protein with a Known Isoelectric Point (Assuming That the Protein Is Only Stable in the pH Range 5.5–8.5)

| Isoelectric point | Ion exchange | Buffer pH |
|---|---|---|
| 8.5 | Cation | $\leq 7.0$ |
| 7.0 | Cation | $\leq 6.0$ |
|  | Anion | $\geq 8.0$ |
| 5.5 | Anion | $\geq 6.5$ |

mM *N*-morpholinoethane sulfonic acid (Mes) buffer pH 6.5 and CM-cellulose will settle this point. If the enzyme adsorbs to either exchanger (or at least disappears from the nonadsorbing fraction), it should at once be treated with 1.0 M KCl. This will almost always be a sufficiently high ionic strength to elute all proteins; then a check is made to ensure that the enzyme is washed off. Sometimes "adsorption" is really "inactivation," so before going ahead on a larger scale the percentage recovery should be checked.

The next stage depends on the result of the previous trial. If the enzyme adsorbed to neither column, a high degree of purification is already indicated. The sample can be passed through columns of both adsorbents in a compromise buffer, e.g., pH 7.0. The two columns can be coupled together; alternatively an intermediate collection and pH adjustment can be used—it would be best not to have to change the buffer itself.

It is particularly convenient to use Tris buffer adjusted to pH 8.0 with Mes for passing the sample through a DEAE-cellulose column. This removes much unwanted protein. Then more Mes (acid form) is used to adjust the nonadsorbed fraction containing the enzyme to pH 6.5 (49). The Mes now becomes the buffering species with $HTris^+$ merely a counterion, and the enzyme sample is immediately run on to CM-cellulose, where in most cases it will be adsorbed.

If the enzyme adsorbed to one of the columns and was eluted successfully by salt, it may be worthwhile seeing whether it would adsorb at a more exacting pH—i.e., higher for CM, lower for DEAE—since this would lessen the quantity of other proteins binding simultaneously. Suitable buffers are described below. Alternatively, an increase in ionic strength at the same pH would have the same effect, e.g., adding 50 mM or 0.1 M KCl to the buffer. A flowchart of progressive trials is suggested in Figure 4.16.

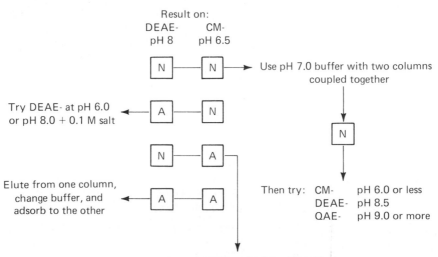

**Figure 4.16.** Flowchart illustrating possible trials of adsorption and elution from ion exchangers. N = not adsorbed, A = adsorbed.

## Buffers for Use in Ion Exchange Chromatography

In all protein work, the correct choice of buffer can be crucial to success. Maintenance of pH in ion exchange chromatography is of paramount importance, since the ionic interactions are so dependent on pH, and the buffers themselves are ionic and may be taking part in the ion exchange process. It is generally advised that the buffering ions should not themselves interact with the adsorbent, that is, that the charged form(s) of buffer should be of the same sign as the substituents on the adsorbent. If this is not so, undesirable and unpredictable pH charges may take place in the microenvironment adjacent to adsorbed proteins, causing, at least, changes in strengths of interaction and, at worst, denaturation. There are, however, many examples in the literature where this rule has been successfully broken; it is less important to observe when higher concentrations of buffer can be used. The commonest nonobservance of the rule is in using phosphate buffers in anion exchange chromatography; sometimes the presence of phosphate is essential for the stability of an enzyme. Wherever possible, one is well advised to follow the rule and allow only simple anions to be present in DEAE experiments (e.g., $Cl^-$, acetate$^-$) and simple cations in CM or other cation ion exchangers (e.g., $K^+$, $Na^+$, $Mg^{2+}$, $HTris^+$ for pH $<$ 7). Remember that polyanionic complexing agents such as EDTA bind to DEAE-cellulose, accumulate on the adsorbent, and may compete with proteins for binding sites.

Often it is necessary to maximize the buffering power at the least possible ionic strength to ensure adsorption of a weakly binding protein. To do this two other rules must be followed. Firstly, the $pK_a$ of the buffer should be not more than 0.5 unit, and preferably not more than 0.3 unit away from the pH being used. Secondly, one of the buffering species should be uncharged, and so not contribute to the ionic strength. For DEAE cellulose at pH 8.0, a Tris-chloride buffer obeys all three rules: the counterion is $Cl^-$, the $pK_a$ of Tris is 8.1 (at 25°C—cf. section 6.1), and the buffering species are $HTris^+$ (noninteractive) and Tris (neutral). For CM-cellulose at pH 6.5, a K-Mes buffer is suitable: the counterion is $K^+$, the $pK_a$ of Mes is 6.2, and the buffering species are HMes (neutral) and Mes$^-$ (noninteractive). On the other hand, phosphate breaks (at least) two of these rules when used with DEAE-cellulose since it is the counterion, and neither buffering form is neutral ($HPO_4^{2-}$ and $H_2PO_4^-$). On CM-cellulose, phosphate breaks the latter rule only.

A list of other possible buffers is given in Chapter 6 (Table 6.3), and further lists are to be found elsewhere (e.g. 50–52).

Suggested buffer compositions for ion exchange use are given in Table 4.6. The strength of the buffer should be at least 5 mM of both ionic and neutral form. Thus when Tris-Cl is used for DEAE-cellulose, the buffer should be no less than 10 mM Tris, adjusted to the $pK_a$ of Tris with HCl. It is convenient to prepare the buffer by starting with 10 mM of HCl and adjusting it to the desired pH with the uncharged base (in this case Tris), thus ensuring that the ionic strength is accurately known (0.01 as $HTris^+$-$Cl^-$). For cation exchangers, the converse strategy is used, e.g., titrating 10 mM KOH with Mes to the

**Table 4.6.** Suitable Buffer Compositions for Ion Exchange Chromatography[a]

| For anion exchange[b] | | | For cation exchange[c] | | |
|---|---|---|---|---|---|
| Temperature (°C) | Buffer | pH | Temperature (°C) | pH | Buffer |
| 20 | Diethanolamine, or 2-amino-2-methyl-1,3-propanediol | 8.5–9.2 | 0–30 | 3.6–4.3 | Lactic acid |
| 0 | | 9.0–9.7 | 0–30 | 4.3–5.2 | Acetic acid[e] |
| 20 | Tris [tris(hydroxymethyl)-aminomethane] | 7.8–8.5 | 0–30 | 4.7–5.4 | Pivalic acid[e] (trimethylacetic acid) |
| 0 | | 8.2–8.9 | | | |
| | | | 0–20 | 5.2–5.8 | Picolinic acid[g] |
| 20 | Triethanolamine | 7.4–8.0 | 20 | 5.8–6.6 | Mes[h] |
| 0 | | 7.8–8.4 | 0 | 6.0–6.8 | Ada[h] |
| 20 | Imidazole[d] | 6.6–7.3 | 20 | 6.3–6.9 | |
| 0 | | 7.0–7.7 | 0 | 6.5–7.1 | |
| 20 | Bis-tris [bis-(2-hydroxyethyl)imino-tris-(hydroxymethylmethane)] | 6.2–6.8 | 20 | 6.8–7.5 | Mops[h] |
| 0 | | 6.5–7.1 | 0 | 7.1–7.8 | |
| 20 | Histidine[d] | 5.6–6.3 | 20 | 7.2–7.8 | Tes[h] |
| 0 | | 5.9–6.6 | 0 | 7.6–8.2 | |
| 20 | Pyridine[ef] | 4.9–5.6 | 20 | 7.8–8.5 | Tricine[h] |
| 0 | | 5.2–5.9 | 0 | 8.2–8.9 | |

[a] All buffers listed are 0.01 ionic strength. The usable pH range extends no more than 0.4 unit either side of the $pK_a$.
[b] 10 mM HCl, adjusted to pH indicated with base form of buffer.
[c] 10 mM KOH, adjusted to pH indicated with acid form of buffer.
[d] Complexes divalent metals, especially histidine.
[e] Volatile.
[f] Poisonous.
[g] UV absorption.
[h] For full names of zwitterionic buffers, see Table 6.3.

appropriate pH. The effect of temperature on $pK_a$ values should be borne in mind (cf. section 6.1), especially if the buffer is made up at room temperature but used in the cold. Unfortunately, when following published procedures, it is rarely clear whether the buffer was adjusted to the stated pH in the cold or at room temperature.

If the enzyme of interest adsorbs strongly at low ionic strength, it may be appropriate to increase the ionic strength, either by adding a neutral salt (KCl, NaCl) or by increasing the buffer strength. The latter choice may be inhibited by the expense of some of the more exotic buffers. A higher buffering power makes the maintenance of exact pH more certain and may enable an elution by change of pH. (cf. section 4.2). (Discussion of pH values and buffers is continued in section 6.1.)

## Conditions for Adsorption

Once an appropriate buffer is chosen, one must consider how to load the column, what the column dimensions ought to be for a given amount of protein, and what other operating procedures are required. It is normal for the applied protein mixture to be in exactly the same buffer used to preequilibrate the column. If this is not necessary, because the enzyme of interest is strongly bound, so much the better, since time is saved not having to transfer the sample into a precisely determined buffer. Nevertheless, the pH should be identical with the column buffer, and the ionic strength preferably similar. There are two usual ways of attaining the correct buffer composition: dialysis or gel filtration. Dialysis is the traditional method, and although it takes a long time, there are few manipulations. More frequently used these days, especially when working on a small scale, is "desalting" by gel filtration. These processes have been described in section 1.4.

The protein sample is now in an appropriate buffer for applying to the ion exchange column. The only remaining adjustment that might be needed is the protein concentration itself. Even after manipulations to remove salts etc., the concentration may be as high as 30 mg ml$^{-1}$, and this is too high, especially if a substantial proportion is going to adsorb to the exchanger. Ion exchange involves buffer ions, and a rapid displacement of counterions by protein molecules could lead to sharp local changes in pH and salt concentrations (cf. section 4.2). Any carefully defined conditions should be reproducible; whereas it is always possible that applying the sample at high protein concentration may work well in a particular case, it may not be readily reproducible. It is the common experience that conditions that are reproducible from day to day are very difficult to genuinely reproduce from laboratory to laboratory, so it is preferable to have methods that are not critically dependent on precise values of any parameter. The more dilute the protein sample is during application, the more ideal and reproducible the result will be. But very dilute solutions are impracticable, and might cause other problems if the $\alpha$ value of the desired enzyme is less than 0.95 (see below). A maximum concentration of adsorbing protein of 5 mg ml$^{-1}$ is allowable, with a further 10 mg ml$^{-1}$ of nonadsorbing protein. At first it may seem strange that any restriction be put on the latter. But remember that nonadsorbing protein carries counterions, which increase the ionic strength, weakening the proper interactions required during sample application.

Usually one assumes that in the conditions for adsorption, the protein will stick to the top of the column and stay there until buffer conditions are altered for the elution procedure. This is so if $\alpha = 1.0$ or very close to that value. But it may not be possible to reach such a state; the pH required may be outside the stability range, or one may have chosen to operate in less strongly adsorptive conditions to lessen the total protein adsorbing to the column. Consider the situation of $\alpha = 0.90$. At any point on the column reached by the protein, 10% remains in solution and passes down to the next point. Moreover, a fraction of the adsorbed protein is constantly desorbing and so being carried by the buffer flow to a lower point. On the average, each protein molecule moves down the column at 10% of the speed of the buffer, so after 10 column volumes of starting buffer have been applied, the protein will begin to emerge from the bottom.

If there are more strongly adsorbing proteins in the mixture being applied, the $\alpha = 0.90$ protein will be displaced more rapidly, since on desorbing from initial binding sites at the top of the column it will find most other sites around it occupied irreversibly and will have to move down with the buffer to an area where these other proteins have not yet reached. In these circumstances it is clear that the applied volume should not be too much greater than the column volume, nor should much washing with starting buffer occur before the elution scheme is commenced. When dealing with complex mixtures of proteins which have some color, it is usual to note a banding at the top of the column as the individual components displace each other in order of partition coefficient. These bands become particularly sharp after all the protein has been applied and one or two column volumes of starting buffer are washed in.

## Size and Dimensions of the Column

The protein concentration and the relative volumes of column and sample size clearly specify the minimum size of column to be used. More important are the actual capacity of the column for the proteins being adsorbed and the length of chromatographic separation needed (sometimes referred to as "equivalent plates"). For some applications, particularly stepwise elutions and affinity elution (cf. section 4.4), a short column can be used with up to one-half of its volume being used for the initial adsorption. As the full chromatographic development is not required in stepwise procedures, it is better to minimize the "unused" part of the column to lessen diffusion. However, since stepwise procedures are rarely true on–off situations ($\alpha = 1.0$ to $\alpha = 0$), partial retention in the "unused" lower part of the column will in fact affect the elution pattern. For more subtle elutions involving gradients a reasonable column length is needed, from 5 to 20 times the amount needed for adsorption.

The adsorptive capacity of ion exchangers for proteins can be very high, but it depends on molecular size. Small proteins bind in amounts in excess of 100 mg cm$^{-3}$, but very large proteins ($>10^6$ MW) cannot penetrate the exchanger and bind only to the surfaces of the particles, resulting in very low capacities (cf. section 4.2) (38,41). Thus it is difficult to predict a priori how much of the adsorbent is involved in the initial uptake of proteins. An average figure for complex mixtures of 30 mg cm$^{-3}$ might be used, but the distribution through the column may well be asymmetric. In addition to the banding effect alluded to above, relative exclusion of larger molecules would cause a large protein to occupy a greater proportion of column than the same amount of a small one. If the molecular size composition of the applied fraction is known (e.g., a fraction from gel filtration, section 5.1), then a better estimate can be made of the likely capacity of the ion exchanger. Using the general figures quoted above, if one has 1 g of protein and expects about half of it to adsorb, it would be reasonable to apply this in 100 ml of buffer to a column of 50 to 100 cm$^3$ vol. Some 15–20 cm$^3$ at the top of the column would be used in adsorbing proteins. These calculations are intended for "strongly adsorbing" proteins, i.e., $\alpha$ values very close to 1.0; if there are proteins with $\alpha$ values above zero but below 0.95 in the mixture, they will be partially retained, and distribute further down the column during the application phase.

In the early days of protein ion exchange chromatography, the dimensions of columns were made very long and thin. Although ideal for true chromatography, the low flow rate necessitated by such a configuration means a long process time, possibly with harmful effects on the protein (cf. section 7.1). Shorter, squat columns with lengths 4–5 times their diameter are more common these days. Theoretically, with perfectly even application and flow through the column, the shape ought to be irrelevant, apart from considerations of particle size affecting the equivalent "plate number" (Figure 4.17; cf. also section 5.1). In practice, perfect flow through a column is difficult to achieve; longer columns even out imperfections better but (for a given linear flow rate) take longer to run. For routine use columns 4–5 times longer than their diameter are recommended. But for stepwise elution ion exchange columns should be still squatter, with heights being only 1–2 times the diameter.

Flow rates must always be a compromise between perfection of chromatographic behavior (at zero flow rate!) and a rough-and-ready separation achieved in minimal time. Modern materials pack so well into columns that they can be operated at up to 100 cm hr$^{-1}$,* much too fast for most procedures. Flow rates of between 10 and 30 cm hr$^{-1}$ are advised. Even so, in some situations rates have to be decreased below this because the column cannot be operated faster, due to large quantities of adsorbed (and sometimes precipitated) protein at the top of the column, and perhaps to viscous solutions of nonglobular proteins. A blockage due to precipitation or denaturation of proteins on

---

*Linear flow rates expressed in cm hr$^{-1}$ can be converted to volume flow rates by multiplying by the cross-sectional area in cm$^2$, giving cm$^3$ hr$^{-1}$.

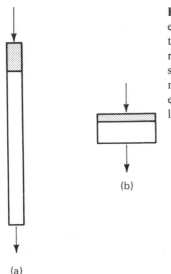

**Figure 4.17.** Columns of identical volume but different proportions. The long, thin column (a) with protein adsorbed to the top 20% has a relatively slow flow rate, but uneven flow is not usually a problem. The squat column (b) can have a very rapid flow rate, meaning a short processing time, but the flow must be even through the whole cross-section, otherwise resolution is lost (also cf. Figure 5.9).

(b)

(a)

the column surface can be temporarily overcome by gently disturbing the surface; nevertheless, blockages should be avoided if possible. Never apply a sample that has a precipitate in it; centrifuge or filter first, in the same conditions of pH, ionic strength, and temperature used for sample application.

## Procedures for Elution of Proteins from Ion Exchangers*

Before describing the different methods of desorbing proteins from ion exchangers, it is useful to repeat (cf. section 4.1) two situations—the trivial one of $\alpha = 0$, where the protein passes straight through the column, and the intermediate situation where $\alpha$ is substantially greater than 0 but significantly less than 1. When $\alpha = 0$, all similar proteins (generally those that have the same charge as the adsorbent) pass through and, apart from boundary effects due to different levels of molecular exclusion, are collected in a volume not substantially larger than that applied. If the enzyme required is in this fraction and most proteins have been held back on the column, this is an excellent method for purifying the enyzme—recovery is usually 100%. Although the full potentialities of chromatography have not been realized, such a procedure of nonadsorption is often used to decrease the total amount of protein being handled.

In the intermediate cases, $\alpha$ from 0.5 to 0.9 (values below 0.5 can be classed with the above, since most of the enzyme will be eluted with the nonadsorbed fraction, unless the sample applied has a very small volume—Figure 4.3), significant retardation of the applied enzyme may lead to complete separation from the nonadsorbed fraction, but elution will occur eventually in the starting

---

*Also cf. section 4.2.

buffer. Two examples of possible results for $\alpha = 0.5$ are shown in Figure 4.18; the separation will depend greatly on the relative volumes of the column and the applied sample. Note that the total volume of buffer that the enzyme is eluted in will be at least as great as the volume of applied sample even if there is no tailing due to a spread of $K_p$ values. In practice the range of $K_p$ (and so $\alpha$) results in elution over an even greater volume. Nevertheless, if there is no problem in handling a large volume of eluted enzyme, this simple method of chromatography without buffer change can give a very good purification for proteins with $0.9 < \alpha < 0.5$.

In most cases the enzyme required is tightly bound in the starting conditions, and a buffer change is needed to elute it. There are three ways of decreasing $\alpha$ values on ion exchangers: (a) by changing the pH (up for cation exchangers, down for anion exchangers), (b) by increasing ionic strength, and (c) by affinity methods. Method (c) is described in detail in section 4.4; (b) is the most commonly employed, and may be used in either stepwise or gradient form. Method (a), pH change, can be useful provided that the buffering power of the new buffer is high enough to titrate both the protein molecules and any adsorbent groups that may become charged or uncharged as the pH changes (cf. section 4.2). For stepwise applications, pH elution can be quite successful; the pH change will be delayed compared with the new buffer front because of these titrations, but eventually a protein peak emerges, coincident with a rapid pH change. But for gradient work, slow pH changes are usually only successful if the buffering power is high—hence high ionic strength (cf. section 4.2).

It is worth remembering the typical titration curve of a protein becomes steep below pH 5.5 (and above pH 9.5), but is very shallow in the pH range 7–8.5. Thus pH gradients need to cover a large range close to neutrality, but very sharp elution may be achieved in a narrow pH range away from neutrality. High pH values are not usually employed because of (a) lack of enzyme

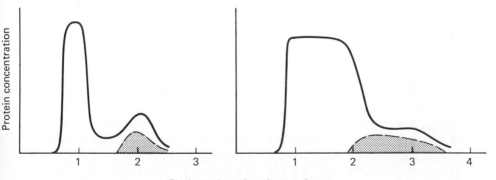

Elution volume in column volumes

**Figure 4.18.** Elution of nonadsorbed ($\alpha = 0$) and retarded ($\alpha = 0.5$, shaded) proteins from an ion exchanger: (a) sample volume $< 0.5$ column volume; (b) sample volume $= 1$ column volume. Note that any tailing of nonadsorbed proteins with $\alpha$ values slightly greater than zero will cause a considerable overlap in (b).

stability above pH 9 and (b) lack of suitable adsorbents: DEAE-adsorbents become uncharged, and few proteins are basic enough to stick to cation exchangers at such high pH values. But some proteins are sufficiently stable in the pH range 4–6 for a lower pH to be used, both DEAE and cation exchangers being suitable depending on the isolectric point of the desired enzyme. (But note the Donnan effect in this regard—Figure 4.11.) A good example of this is the separation of the two isoenzymes of yeast hexokinase on DEAE-cellulose in the pH range 5.6–5.2 (53). Many aspartate and glutamate residues become protonated as the pH falls by only a fraction of a pH unit; consequently the gradient, in pH terms, can be very gentle.

Ionic strength gradients do not have any problems of irregular behavior, and increasing salt strength should emerge from the bottom of the column in the same form as the gradient was applied at the top. The salt concentration needed to elute proteins will depend largely on isoelectric points, but few proteins remain reversibly adsorbed to ion exchangers above $I = 0.5$; some require 1 M salt to remove them, but rarely more. It is more common to need high salt concentrations with DEAE-cellulose, since isoelectric points below the $pK_a$ value for aspartate/glutamate residues ($<$pH 4.5–5) are much commoner than isoelectric points above $pK_a$ values for lysine ($>$pH 10). Net charges (at neutrality) of low-isoelectric-point proteins can be quite high, but net charges of high-isoelectric-point proteins are usually less. It is rare for any proteins to remain adsorbed to CM-cellulose at neutral pH in the presence of 0.2 M salt. Phosphocellulose is different, and behaves somewhat in the fashion of an affinity adsorbent; quite high salt concentrations are often needed to break the pseudospecific binding.

Gradients (of ionic strength or pH) can be formed using a simple apparatus commercially available in which two containers are connected at the bottom, and an efficient mixing paddle is in the container from which the buffer is extracted. Simpler still is the two-beaker method with a magnetic flea to stir the sampled buffer (Figure 4.19). Much more complex, and mainly useful for routine, frequently repeated procedures rather than for developing new preparative methods, are electronically controlled mixers which can produce any gradient profile required. Fortunately this apparatus is not essential for the protein laboratory, as it costs some three orders of magnitude more than the two-beaker method! But it does dispense with two pieces of apparatus that are

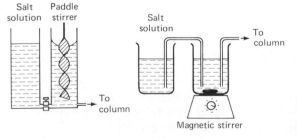

**Figure 4.19.** Simple gradient mixers (linear gradients).

needed with the beakers, namely the magnetic stirrer and the pump. By arranging the gradient mixing device above the column it is possible to run the column under gravity feed, but a peristaltic pump to deliver a constant flow is preferable. Pumps come in many shapes and sizes and, above all, costs. An expensive pump should be capable of a continuously variable flow rate, minimum pulsation, and capability of operating against a fair pressure without decreasing the flow rate. However, many of these capabilities are not always necessary—in particular the latter, since it may be better for the pump to cease flowing against an increased pressure due to a blockage than to burst the tubing and continue to pump liquids all over the bench. In general pressure should *not* increase during the run—if it does, there is probably something wrong. Besides the excellent pumps available from many major companies specializing in biochemical equipment (LKB, Pharmacia, Ismatec), there are simpler and much cheaper versions which are adequate for most purposes.

Finally, here are some comments on stepwise salt elution. Compact peaks containing the enzyme can be obtained by a stepwise application of a buffer with higher ionic strength, and this can be a rapid and efficient way of purifying an enzyme. However, because of other factors affecting the $\alpha$ value, namely protein concentration and weak and strong binding sites (cf. section 4.1), it is unlikely that a simple on–off system will operate well unless the step in salt concentration is quite large. With a large increase in salt concentration, many other proteins may be eluted in addition to the desired one. A sudden decrease in the $\alpha$ value due to the salt brings off much of the enzyme, but tailing effects are large, and a further increase in salt may be needed to bring off the rest of the enzyme (Figure 4.20). The second peak is *not* a different enzyme form; it would be most unwise to deduce the occurrence of isoenzymic

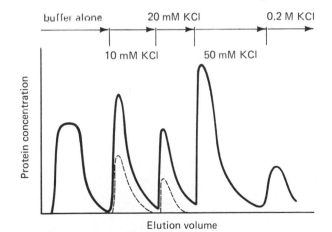

**Figure 4.20.** Stepwise elution from ion exchange column by increasing salt concentration. Enzyme (dotted line) has been distributed between two peaks, but the enzyme does not necessarily exist in two forms. Note the sharp leading edges and long trailing edges to the eluted protein peaks (see text). [From Scopes (29).]

forms from a stepwise elution procedure. The tailing of a peak is lessened by having a gradient—indeed, this is one of the chief reasons for using gradient elution so extensively (Figure 4.15).

A few further points are worth noting about ion exchange chromatography. Whereas simple theories imply that conditions for elution are exactly the inverse of those for adsorption, in practice some hysteretic behavior is often noted. This is particularly so with phosphocellulose, and has also been noted with other adsorbents such as affinity- and dye-ligand adsorbents (cf. section 4.5 and 4.6). Hysteresis means, for example, that the $\alpha$ value operating when a particular ionic strength buffer is used for adsorption will be different from the $\alpha$ value when the same buffer is used for elution. Thus in Figure 4.21 ionic strength A may be suitable for adsorption, and ionic strength B will not cause elution. But ionic strength B would not be suitable for adsorption. These effects must be explained in terms of nonideality, nonequilibrium processes, and other factors not taken into account by simple theory.

Another observation which is very important in cation exchange chromatography concerns the influence of other charged polymers, in particular nucleic acids, present in the sample being applied. Electrostatic attractions between charged macromolecules can be strong at the low ionic strengths used in ion exchange chromatography, and positively charged proteins interact with negatively charged nucleic acids. The complex formed is less positive, or perhaps negative overall, and so may not bind to a cation exchanger as expected. A simple remedy to avoid this problem is to treat the sample with a polycation (usually protamine) which specifically causes precipitation of nucleic acids, with little protein being removed. Protamine sulfate, used at a level of about 1 g for every 20–40 g protein, should be first dissolved in water and adjusted to neutrality (it is strongly acid) before adding to the sample. The ionic strength

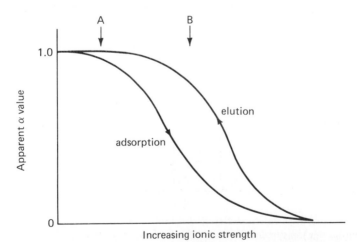

**Figure 4.21.** Hysteresis in ion exchange chromatography. Due to nonequilibrium operation, conditions for adsorption appear to be slightly different and more demanding than conditions for elution (see text).

should not be high ($<0.1$); protamine–nucleic acid complexes are also a result of electrostatic interactions and are dispersed by high salt concentration. Provided that sufficient protamine has been used, the thick creamy white precipitate can be removed by moderate-speed centrifugation. Many times have people concluded that cation exchange chromatography is no use for their enzyme because it "does not stick"—and the reason has been interference by nucleic acids.

## 4.4 Affinity Elution from Ion Exchangers and Other Nonspecific Adsorbents

Affinity chromatography to most people implies the use of an immobilized ligand on an adsorbent which specifically selects out proteins binding to that ligand (cf. section 4.5). After adsorption, the enzyme required can either be nonspecifically eluted by, for instance, an increase in salt concentration or specifically eluted by displacement with ligand free in solution. Thus affinity chromatography can involve two specific stages, "affinity adsorption" and "affinity elution." The mechanism by which affinity elution from a specific adsorbent operates is clear. In practical terms the presence of a ligand reduces the partition coefficient from near 1.0 to a significantly lower value. However, the latter could occur with other adsorbents; the term *affinity elution* does not require any particular property of the adsorbent itself. Affinity elution from ion exchangers has been found to be a very convenient and specific method for purifying many enzymes, where they have a certain combination of properties (54–58).

The basic principles of affinity elution from ion exchangers are illustrated in Figures 4.22, 4.23, and 4.24. An enzyme is adsorbed to an ion exchanger at

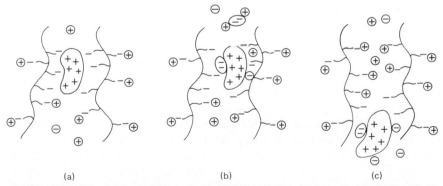

(a)                              (b)                              (c)

**Figure 4.22.** Principles of affinity elution from cation exchanger. In (a) the positively charged protein is adsorbed to the exchanger. In (b) a negatively charged ligand that binds to the protein is introduced, lessening the total charge on the protein. In (c) the protein is eluted, since $\alpha$ has decreased due to the lower total charge on the protein.

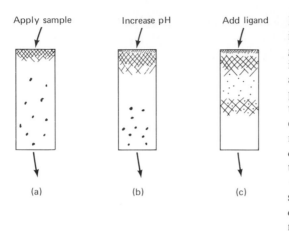

Apply sample          Increase pH          Add ligand

(a)                        (b)                        (c)

Figure 4.23. Principles of affinity elution from cation exchanger. In (a) the protein mixture is applied; the enzyme required adsorbs, some other proteins pass through the column. In (b) the pH is increased so that the enzyme required begins to move down the column (0.8 < $\alpha$ < 0.95)—some other proteins may be totally eluted ($\alpha$ < 0.8). In (c) the addition of specific ligand decreases $\alpha$ (0 < $\alpha$ < 0.5) so that the enzyme required is eluted.

a certain pH because the electrostatic interactions between the matrix of the adsorbent and the charges on the protein are sufficiently strong to hold it, with a partition coefficient close to 1.0. If a ligand of the enzyme is introduced which is charged and of opposite sign to the net charge on the enzyme, then binding of this ligand decreases the net charge on the enzyme. In the right conditions the reduction of the net charge on the enzyme weakens its electrostatic interaction with the adsorbent sufficiently to reduce the partition coefficient substantially. This might result in a reduction in from 0.95 (when the enzyme band moves slowly down the column at a rate of one-twentieth that of the buffer flow) to 0.5 (the enzyme moves down at one-half the rate of the buffer). This causes specific elution of that enzyme and no other—except those which might also bind that ligand *and* have similar adsorption characteristics. The first proteins for which this method was successfully used were fructose 1,6-bisphosphate aldolase and fructose 1,6-bisphosphatase (54,55). In both cases the tetrameric enzyme (of about 150,000 MW) bound 4 substrate molecules having 4 negative charges each. Consequently a reduction in total (positive) charge of 16 units was able to cause a large diminution in adsorption (Figure 4.25). Calculations based on experimental values (38) show that for rabbit aldolase on CM-cellulose at pH 7.1 the charge addition should reduce $\alpha$ from 0.95 to 0.01; even if adsorbed much more strongly initially, at $\alpha$ = 0.995 the addition of substrate would reduce the value to 0.35. However, many enzymes do not bind so many molecules of substrate, nor are their substrates as charged, so the total effect might in general be expected to be weaker than for aldolase.

A detailed study on the adsorption of various enzymes to CM-cellulose in the presence and absence of substrates demonstrated that the effect was in fact larger than expected in some cases—$\alpha$/pH curves were shifted to lower pH in the presence of ligand (38). *At a given $\alpha$ value* the total charge on the enzyme and on the enzyme-ligand complex could be calculated, using an enzyme titration curve to relate the two pH values. With some enzymes it was found that the shift corresponded to no more than the change in charge expected by titrating the enzyme through that pH range. For example, the $\alpha$/pH curve for rabbit

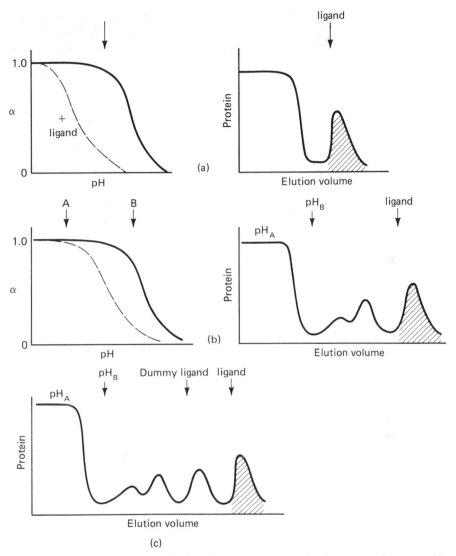

**Figure 4.24.** Operation of an affinity elution column. (a) Application at the same pH as used for elution; the effect of the ligand is so large that $\alpha$ decreases from $>0.95$ to a very low value. (b) Application at $pH_A$ where $\alpha \simeq 1.0$, and increase of pH to B to make $\alpha \simeq 0.9$, so that addition of ligand has sufficient effect (as in Figure 4.23). (c) As (b), but inclusion of "dummy ligand." In practice the "dummy ligand" can be used at the commencement of $pH_B$ wash. Shaded area represents the enzyme being eluted.

muscle pyruvate kinase shifted only 0.25 pH unit in the presence pf 0.1 mM phosphoenol pyruvate. This tetrameric enzyme is expected to acquire 12 negative charges if it becomes saturated with its substrate. However, it would lose only about 8 positive charges by titrating through 0.25 pH unit. In this instance, 0.1 mM phosphoenol pyruvate was probably insufficient to saturate

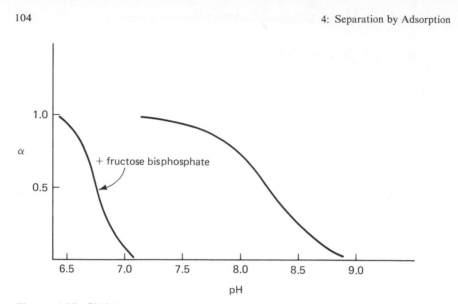

**Figure 4.25.** Shift in $\alpha$ against pH curve for rabbit aldolase in the presence of 0.1 mM fructose 1,6-bisphosphate (FBP). [From Scopes (38).]

the enzyme. On the other hand, lactate dehydrogenase, on binding 4 molecules of NADH, would acquire 8 negative charges, corresponding to a pH shift of about 0.8 pH unit (in the range pH 8.5–7.0). But in fact the addition of 0.1 mM NADH shifted the curve by up to 1.5 pH units—nearly twice the effect expected. Similarly with yeast phosphoglycerate kinase (48), the shifts on binding 3-phosphoglycerate or ATP were rather more than anticipated, and with 1,3-bisphosphoglycerate substantially more.

Two other explanations for the affinity elution effect can be proposed. Besides simply canceling out total charge, the ligand binds at a site which probably has a cluster of charges opposite in sign to those of the ligand. This binding may mask a rather specific strong interaction between these charges and the adsorbent. Also, since the active site is likely to be close to the surface of the protein molecule, charged residues at the active site would be having more influence on the total interaction than other charges further from the surface. The masking of these charges with ligand would weaken the interaction more than would the loss of a similar number of charges randomly distributed through the protein molecule.

Alternatively, a protein conformational change may occur when the ligand binds. It is known that conformational changes do occur (59,60), and this could well alter the interaction with adsorbent—in either direction. Convincing evidence of a conformation change that results in *tighter* adsorption has not been found though the affinity elution effects with some enzymes (notably yeast enolase) have been so slight, less than the expected charge effect, that an opposing conformation effect may have been operating. But *weaker* adsorption due to conformational changes does seem to occur, notably with phosphoglycerate kinase, adenylate kinase, and lactate dehydrogenase (38,48).

In general, it can be said that an efficient affinity elution procedure can be found if the following characteristics of the enzyme apply:

(a) The enzyme can be adsorbed, without inactivation, on an ion exchanger which has the same sign of immobilized charge as the ligand to be used in affinity elution. For anion exchangers (DEAE etc.), the ligand must be positively charged; for cation exchangers (CM, phospho, etc.), the ligand must be negatively charged.
(b) The ligand must be capable of binding to the enzyme at a pH where the enzyme otherwise remains on the column (in practice with a minimum $\alpha$ value of 0.9).
(c) The number of charges added by the ligand per $10^5$ MW of enzyme should be at least 4, unless an additional conformational charge assists the elution.
(d) The dissociation constant for the ligand should preferably be below $10^{-3}$ M; the smaller the better.

In view of restriction (a) above, since practically all ligands (other than metal ions) are negatively charged, affinity elution from ion exchangers is virtually restricted to proteins of a neutral or basic nature that will adsorb to a cation exchanger at neutral pH (6–8). The affinity elution effect does in fact work with $Mg^{2+}$ ions for pyruvate kinase adsorbed to QAE-Sephadex at high pH (Scopes, unpublished observations), but it is very weak, owing to restrictions (c) and (d). It is possible that enzymes such as arginine kinases, which are purified using DEAE-cellulose, might benefit using affinity elution (with arginine), though again restrictions (c) and (d) would make the effect weak. Polycationic ligands are rarely encountered. Many animal enzymes have isoelectric points in the appropriate range for cation exchange affinity elution (> pH 6), but corresponding enzymes in plants or microorganisms often have lower isoelectric points. Since the methods depend quite critically on isoelectric points, they are species-specific; changing species may require a reexamination of the appropriate buffer pH values for optimum procedure.

## Practical Steps in Affinity Elution

The following describes the stages in an affinity elution purification. The sample is equilibrated with an appropriate buffer and applied to an ion exchange column—assume it is CM-cellulose. The required enzyme all adsorbs; the sample is washed in with a little starting buffer. If the affinity elution effect is considerable, then the starting buffer pH can be chosen to be the same as the elution pH. More generally, it will be necessary to raise the column pH to a suitable value, at which the enzyme required is just beginning to move, e.g., $\alpha$ = 0.95–0.90 (Figures 4.23 and 4.24). The ligand is then applied at a concentration sufficient to exceed the dissociation constant for the enzyme–ligand complex, and if there is a large amount of enzyme protein present, the total *amount* of ligand in molar terms must exceed the molar *amount* of binding sites. On the other hand, the concentration of the ligand should not make a substantial difference to the ionic strength of the buffer. Expense often dictates the maximum amount of ligand that is used, but if its dissociation constant is rather high and the addition of a sufficient concentration of ligand causes a significant increase in the ionic strength of the buffer, then a "dummy ligand wash" may be needed. In this method, a compound of similar charge characteristics to the ligand, but which does not bind to the enzyme, is used in the preelution wash. Any proteins that are removed from the column by the increase in ionic strength are taken out at this stage; the column is now ready for affinity elution, and the "dummy" ligand is replaced by the true one (Figure 4.24c).

For anion exchangers, the sample is applied at a higher pH and eluted after reducing the pH. But remember that when using anion exchangers, the ligands must be *positively* charged. If a negatively charged ligand is added, complex effects occur; most likely the ligand itself will displace proteins unspecifically by a conventional ion exchange procedure—the ligand will bind to the exchanger more tightly than the proteins. It will also bind specifically to its enzyme, but (at least theoretically) this should make the enzyme adsorb even more tightly, since it acquires more negative charges. A possible "negative affinity elution" has been attempted in which the enzyme is bound, washed at a critical pH in the presence of its ligand, and then washed off in the absence of the ligand. If the ionic conditions are such that the ligand is not substantially adsorbed to the exchanger, this can work; the ligand should have no more than 1 negative charge, or else it is almost certain to bind to the exchanger (Figure 4.26).

## Some Theoretical Calculations

On the basis of experimental data obtained with purified enzymes (38), some calculations can be made, and compared with the data, assuming equilibrium binding conditions are established. Developing the ideas presented in section 4.1, the following equilibria should be pertinent for a protein $P$ that binds $n$ ligands $L$ in the presence of an adsorbent with an effective binding site concentration $M$:

$$P + M \rightleftharpoons PM \qquad K_P$$
$$P + L \rightleftharpoons PL \qquad K_L^1$$
$$PL + L \rightleftharpoons PL_2 \qquad K_L^2$$
$$\cdot$$
$$\cdot$$
$$\cdot$$
$$PL_{n-1} + L \rightleftharpoons PL_n \qquad K_L^n$$

$$PL + M \rightleftharpoons PLM \qquad K_P^1$$
$$PL_2 + M \rightleftharpoons PL_2M \qquad K_P^2$$
$$\cdot$$
$$\cdot$$
$$\cdot$$
$$PL_n + M \rightleftharpoons PL_nM \qquad K_P^n$$

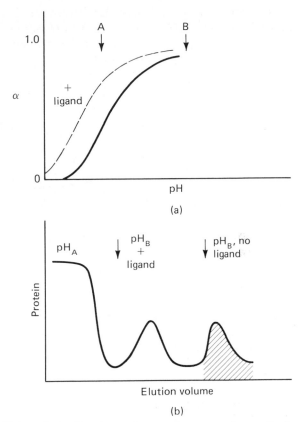

**Figure 4.26.** "Negative affinity elution" using anion exchanger and negatively charged ligand; (a) $\alpha$/pH curves $\pm$ ligand; (b) the elution profile. After adsorbing at $pH_A$, the pH is decreased in the presence of ligand. Nonspecific proteins with low $\alpha$ values at $pH_B$ are eluted. The ligand is then removed from the buffer, whereupon the $\alpha$ value for the desired enzyme decreases [see (a)] and the enzyme is eluted (shaded area). Note that the ligand must not bind to the adsorbent in the conditions used.

By definition of the system, $K_P^n > K_P^{n-1} > \cdots > K_P^1 > K_P$; i.e., each binding of ligand weakens the interaction. Because it is virtually impossible to determine intermediate values of $K_P^i$, only two cases can reasonably be considered. Firstly, where $n = 1$, we can measure $K_P^1$ from the $\alpha$ value at saturating levels of $L$ (Figure 4.27). This is the example of a monovalent enzyme:

$$P + M = PM \qquad K_P$$
$$P + L = PL \qquad K_L$$
$$PL + M = PLM \qquad K_P^1$$

and

$$\alpha = \frac{[PL] + [PLM]}{P_t}$$

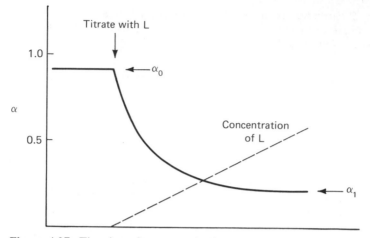

**Figure 4.27.** Titration of enzyme-adsorbent mixture with ligand and determination of $\alpha$ (for methods see Ref. 38). $K_p$ can be calculated using Eq. (4.6) from the value of $\alpha_0$ and $m_t$. $K_p'$ can be calculated from $\alpha_1$ and $m_t$.

is the solution of a cubic equation, using known values of $K_P$, $K_L$, $K_P^1$, $\ell$ (= concentration of $L$), $m_t$, and $p_t$. Alternatively, knowing $K_P$, $K_P^1$ (from $\alpha_0$, $\alpha_1$, Figure 4.27), $\ell$, $m_t$, and $p_t$, it is possible to determine $K_L$, the dissociation constant for the enzyme–ligand complex, from the variation of $\alpha$ with $\ell$. Note that $\ell$ is the concentration of free ligand, not the total concentration present. So if an enzyme has only one binding site for ligand, and the value of $\alpha$ is measured at a fixed pH with varying $\ell$ (as in Figure 4.27), $K_L$ can be measured. However, complications arise when it is remembered that $K_P$ and $K_P^1$ are not unique values, but rather an average of a continuum of values for those particular conditions.

   If there is more than one binding site, then although analysis of the $\alpha/L$ curve could theoretically allow determination of individual dissociation constants for the partially saturated forms, it is improbable that any accurate values could be arrived at. But it is possible to make some calculations for the second case, in which all $K_P^i = \infty$. These show how the displacement of protein from adsorbent depends on the concentration of ligand relative to its dissociation constant, and particularly how much more effective multiple binding sites are in achieving affinity elution at a low concentration of (free) ligand. Multiple binding sites result in more ligand binding to the protein, but on balance the *total amount* of ligand to achieve a given reduction in $\alpha$ is less. Consequently, to the previous list of characteristics required for successful affinity elution can be added the desirable (though not essential) one of multiple binding sites.

   The equation for $\alpha$ in the absence of ligand, in terms of $p_t$, $m_t$, and $K_P$, was given previously as Eq. (4.5):

$$p_t\alpha^2 - (p_t + m_t + K_P)\alpha + m_t = 0$$

If ligand $L$ (concentration $\ell$, dissociation constant $K_L$), when it binds to the protein displaces it totally from the adsorbent (i.e., $K_P^i = \infty$), then the new $\alpha$ value at subsaturating levels of $L$ is given by

$$p_t \alpha^2 - \left[ p_t + m_t + K_P \left( 1 + \frac{\ell}{K_L} \right) \right] + m_t = 0 \qquad (4.9)$$

for a monomer, and

$$p_t \alpha^2 - \left\{ p_t + m_t + K_P \left[ 1 + \prod_{i=1}^{n} \left( \frac{\ell}{K_L^i} \right) \right] \right\} + m_t = 0 \qquad (4.10)$$

for an n-mer, if *all* values of $K_P^i$ are infinite. Now $\alpha/\mathrm{pH}$ curves calculated for various conditions are given in Figures 4.28 and 4.29. It will be noted that the curves are shifted to low pH, but the *shapes* of the curves remain similar. This applies for multiple-binding-site proteins as well as monomers. For the typical tetramer, the pH shift becomes substantial as soon as $\ell$ exceeds $K_L$ because the expression $[1 + \Pi_{i=1}^{n} (\ell/K_L^i)]$ rises rapidly.

In practice, as the pH decreases, the values of $K_P^i$ are not infinite, and so a more complex analysis is needed. But qualitatively one can see that a non-infinite value of $K_P^i$ results in more adsorption, so it is common for actual $\alpha$ values to be higher than those theoretically predicted in Figure 4.28 in the *lower pH range* of the curve. Consequently the real $\alpha$ value increases more steeply in response to pH in the presence of ligand than shown in Figure 4.23. Experimental results with lactate dehydrogenase are shown in Figure 4.30, with a corresponding theoretical curve for $K_P^i = \infty$ drawn in.

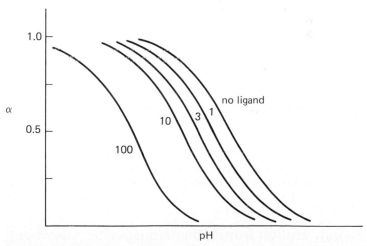

**Figure 4.28.** Variation in $\alpha/\mathrm{pH}$ curves (for a cation exchanger) in the presence of ligand, calculated from Eq. (4.9) for a monovalent enzyme–ligand interaction. It has been assumed that (i) charge addition alone is responsible for the ligand-induced shift, and that (ii) the net charge on the protein is linearly related to pH over this range. Curves from right to left represent concentrations of ligand equal to 0, 1, 3, 10, and 100 times $K_L$, respectively.

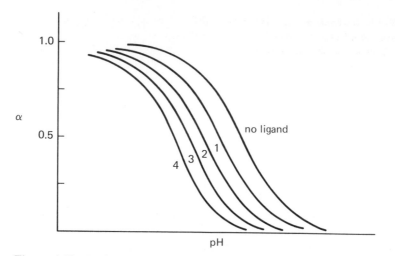

**Figure 4.29.** As for Figure 4.28, but illustrating the effect for multimeric enzymes. Curves 1, 2, 3, and 4 show $\alpha$ values for enzymes with 1, 2, 3 and 4 binding sites for ligand $L$ respectively, when ligand concentration equals 2 times $K_L^i$ in each case.

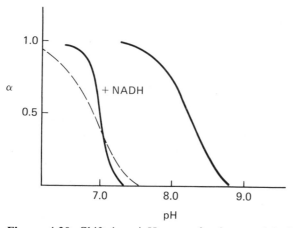

**Figure 4.30.** Shift in $\alpha$/pH curve for lactate dehydrogenase in the presence of NADH. In this case some of the shift is due to conformational change as well as charge addition. Theoretical (dotted) line for $K_p^i = \infty$ has been included. [From Scopes (38).]

## Affinity Elution from Other Nonspecific Adsorbents

Phosphocellulose is an ion exchanger, and many of the published affinity elution procedures use this adsorbent. However, there are specific effects involved with phosphocellulose, in particular with enzymes that bind phosphorylated substrates. Nucleic acid enzymes have usually been purified by adsorption to P-cellulose, and several tRNA synthetases have been successfully eluted using RNA substrates (61). Glucose 6-$P$ dehydrogenase (63), hexokinase (63), and

**Figure 4.31.** Phosphorylated nonreducing end of phosphocellulose as an affinity ligand for glucose 6-phosphate dehydrogenase and phosphoglucose isomerase.

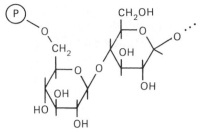

phosphoglucose isomerase (56) also can bind to and be specifically eluted from phosphocellulose; in these cases one may consider that phosphorylated nonreducing ends of the cellulose may be acting in a specific fashion (Figure 4.31). But in most other cases phosphocellulose should be considered simply as an ion exchanger, and discussion of its behavior belongs to the previous section.

Yon (64,65) introduced the concept of multifunctional adsorbents, containing a mixed population of hydrophobic, positively, and negatively charged groups, covalently attached to agarose:

$$-O-C-NH-(CH_2)_n-NH-CO-(CH_2)_n-COO^-$$
$$\underset{^+NH_2}{\overset{\|}{\phantom{x}}}$$

where $n = 8$ or 10.

These were originally developed as spacer arms for affinity adsorbents, but were found to have "nonspecific" adsorptive properties due to their multifunctional nature. Biospecific elution from these adsorbents, named "imphilytes," has been found successful in a number of cases (66,67). Interpretation of the adsorption and elution is much more complex than with ion exchangers, since the actual nature of the protein binding is not at all clear. Although potentially useful, the technique has largely been supplanted by dye-ligand adsorbents which also involve hydrophobic and charged interactions (cf. section 3.6). A major problem with the imphilytes is their low adsorptive capacity; as little as 100 $\mu$g of protein per cubic centimeter of adsorbent was recommended (65).

Porath (68,69) introduced dipolar ion exchangers (also cf. section 4.8), which have high adsorbent capacities and behave somewhat differently from ion exchangers; they have both positively and negatively charged groups. A reinvestigation of these adsorbents for affinity elution methods would seem justified.

# 4.5 Affinity Adsorption Chromatography

In order to avoid confusion with the previous section, this one, which describes what is commonly understood to be "affinity chromatography," is headed "Affinity Adsorption Chromatography." In this case it is at the adsorption stage that the important biospecific selection takes place, where the protein being isolated has a *specific* affinity for the adsorbent, related to its ligand bind-

ing properties (Figure 4.32). There is a blurred division between the truly specific affinity adsorbents which bind little other than proteins that have binding
sites for the immobilized ligand and the general adsorbents which, although
showing specificity for certain classes of proteins, bind many others also. In the
latter class one can commence with the dye adsorbents (cf. section 4.6) which
have a substantial selectivity for nucleotide binding sites on enzymes, and move
down to adsorbents such as phosphocellulose, which, while being principally an
ion exchanger, has often been used as a pseudo-affinity matrix for such things
as nucleic-acid-binding proteins (70) and enzymes interacting with phosphorylated sugars (62,63). Although there are some ready-made insoluble materials which mimic substrates or are actual substrates, e.g., starch (for amylases,
glycogen-metabolizing enzymes), cellulose (cellulases), and phosphocellulose
(see above), affinity adsorbents must otherwise be synthesized by covalently
attaching the ligand to a suitable matrix. The choice of matrix is dictated by
the same sorts of consideration as apply for ion exchangers—namely, it must
be a porous, hydrophilic polymer which can be manufactured in suitable particulate form, allowing free access of macromolecules to the attached ligands,
and an adequate buffer flow when packed in a column. Although other adsorbents have been used and in individual cases might have advantages, spherical
agarose beads with molecular exclusion size about $10^7$ daltons for globular proteins have been used most extensively. Cross-linked agarose beads are becoming more widely used because of their dimensional stability under pressure or
changing salt-solvent conditions. Products such as Sepharose-4B and -CL6B
(Pharmacia) and agaroses from Bio-Rad are usually used.

The procedures used for attaching ligands to the matrix will only briefly be
touched on here; recent publications exclusively concerned with affinity adsorption techniques cover the subject in greater detail (71–73). The main requirements for a successful affinity adsorbent are:

1. The ligand should be attached to the matrix in such a way that its binding
   to the protein concerned is not seriously disturbed.

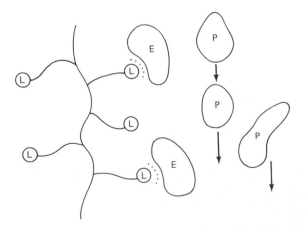

Figure 4.32. Basic principle
of affinity adsorption chromatography. A ligand $L$ is
covalently attached to the
backbone matrix. Only
enzymes $E$ with a specific
affinity for $L$ bind to the
adsorbent. Proteins $P$ pass
through unaffected.

2. A "spacer arm" setting the ligand away from the matrix should be used to make it more accessible to the protein (or for other reasons—see below).
3. Nonspecific interactions should not be so great that many other proteins are adsorbed in addition to the one required.
4. The linkages should be stable to the likely conditions to be used during the chromatography, including "cleaning-up" procedures before re-use.

## Synthesis of Affinity Adsorbents

Methods for synthesizing an affinity adsorbent are based on one of two general procedures: either the ligand or ligand analogue is reacted with adsorbent already possessing the spacer arm, or a ligand analogue containing the spacer arm is reacted with activated adsorbent (Figure 4.33). The spacer arm must have reactive ends for attachment to both matrix and ligand. In some cases the ligand–spacer arm combination is synthesized from smaller components rather than an attempt made to link them in one step, e.g., $N^6$-aminohexyl AMP (74).

The chemistry of the activation process involves the reaction of hydroxyl groups on the agarose matrix, and the most commonly used reaction has been

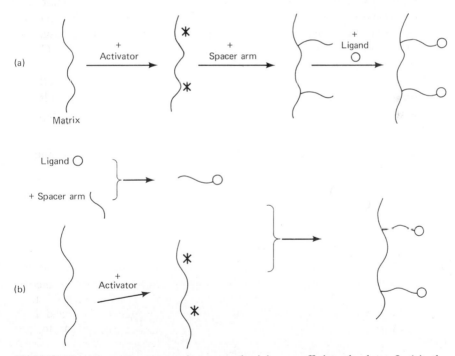

**Figure 4.33.** Alternative approaches to synthesizing an affinity adsorbent. In (a), the spacer arm is attached to the matrix, then the ligand is covalently attached. In (b) the spacer arm is attached to the ligand first, and this is then reacted to link up with the matrix.

with cyanogen bromide. The original method (75) has been superseded by a simpler technique in which the reaction is complete in a few minutes (76), Cyanogen-activated agarose available commercially has been widely used. The reactions of cyanogen bromide are complex; both (i) cyclic (Figure 4.34) and (ii) acyclic imidocarbonates are formed, with (iii) some carbamates and (iv) carbonates as side products, but the principal reactive component has been demonstrated to be (v) the cyanate ester (77). Indeed, the imidocarbonates, being unstable in acid, are destroyed by an acid treatment in the Pharmacia preparation of cyanogen–bromide-activated Sepharose.

The activated agarose reacts swiftly in weakly alkaline conditions (pH 9–10) with primary amines to give principally the isourea derivative (Figure 4.35). Some simultaneous aqueous hydrolysis gives carbamates (iii) and carbonates (iv) as in Figure 4.34.

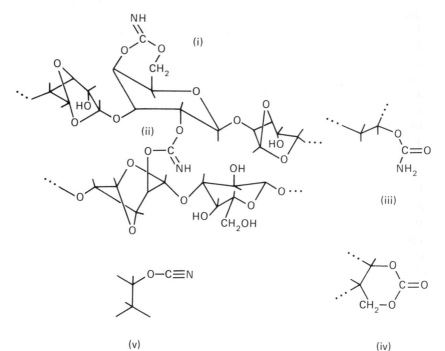

**Figure 4.34.** Reaction products of cyanogen–bromide with agarose: (i) cyclic, and (ii) acyclic imidocarbonates, (iii) carbamate, (iv) carbonate, (v) cyanate ester.

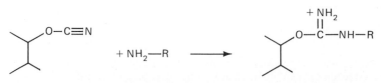

**Figure 4.35.** Reaction of cyanate ester (Figure 4.34v) with amino compound to give isourea derivatives.

The isourea substituent is positively charged at neutral pH, which can influence the behavior of affinity adsorbents by introducing an element of anion
exchange character. On the other hand, many of the ligands that are attached
are negatively charged, so the isourea derivative may cancel out possible cation
exchange effects, though leaving dipolar ion characteristics (69).

Although cyanogen bromide activation is still the most widely used method,
especially for attachment of proteins through ε-lysine groups (cf. section 4.7),
other more convenient and efficient methods are now available and undoubtedly will eventually supplant the cyanogen bromide procedure. The first of
these is "epoxy activation," in which an epoxy group is introduced, using 1,4-
butanediol diglycidyl ether (a bisoxirane) (Figure 4.36) (78).

In the latter case a spacer arm is automatically inserted, since there are 11
atoms between the matrix and the reactive epoxy group. Epoxy groups are
somewhat more reactive than cyanates, and as well as combining with primary
amines, they react with hydroxyls to form ether linkages (Figure 4.37). The
coupling is carried out at pH 8.5–11; the reaction with hydroxyls is slow, taking
about 24 hr at room temperature.

The third method involves activation with 1,1'-carbonyldiimidazole, which
gives a product just as reactive as the cyanate; coupling (again to a primary
amine) does not leave a positive charge, but a urethane linkage (Figure 4.38).
The reagent is also less noxious than either cyanogen bromide or bisoxiranes.

The most recent method, which may well be the best, involves the use of
toluene sulfonyl chloride (tosyl chloride) (80), or the more reactive 3,3,3-tri-
fluoroethanesulfonyl chloride (Tresyl chloride) (81). The tosyl group introduced on activation reacts smoothly and rapidly to give the very stable secondary amine from the primary amine ligand (Figure 4.39). At neutral pH the
secondary amine group is charged, so the ultimate behavior is very similar to
that of cyanogen bromide activation, but the linkage is much more stable.

In practice, $RNH_2$ in the examples above is usually either the ligand with
spacer arm ending with the amine already attached, or hexamethylenediamine,
which gives a spacer arm ending with another primary amine; i.e.

$$\rangle^* + NH_2(CH_2)_6NH_2 \rightarrow \rangle - NH(CH_2)_6NH_2$$

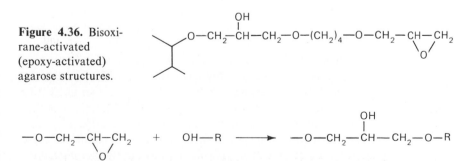

**Figure 4.36.** Bisoxi-
rane-activated
(epoxy-activated)
agarose structures.

**Figure 4.37.** Reaction of epoxy-activated agarose with hydroxyl residue.

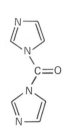
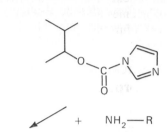

+

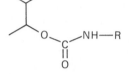

+     NH<sub>2</sub>—R

**Figure 4.38.** Activation and reaction of carbonyldiimidazole-activated agarose.

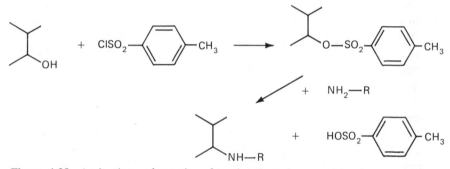

**Figure 4.39.** Activation and reaction of tosyl-activated agarose.

The further chemistry of attachment of particular types of ligands and spacer arms is beyond the scope of this section, but is extensively reviewed elsewhere.

Many affinity adsorbent matrices, either "activated" or with spacer arms or complete, are available commercially. The complete adsorbents mostly contain ligands which are not specific for one enzyme, but are general ligands which bind to many different proteins. In particular, nucleotide ligands have been extensively investigated since so many enzymes use nucleotides as cofactors or substrates. Thus although the specificity of these adsorbents is far from absolute, they are generally useful, being applicable to a range of different enzymes. It has been stated that 31% of all known enzymes utilize one of four nucleotide cofactors (82); if one includes nucleic acids as substrates, and enzymes that bind nucleotides as activators or inhibitors, the total exceeds 50%. This does not necessarily mean that 50% of the protein in a given extract consists of these enzymes, nor that many of them will necessarily bind to a particular nucleotide affinity adsorbent in specified conditions. A list of commercially available adsorbents as of 1978 has been presented elsewhere (72); catalogs by the major companies (P-L Biochemicals, Sigma, Pharmacia, Bio-Rad) extend this list much further.

## Application of Chromatographic Theory to Affinity Adsorption

The theory developed in section 4.1 can easily be applied to affinity adsorption chromatography, with some interesting conclusions. Since the value of $K_P$, the dissociation constant between adsorbent and protein, should be similar or related to the dissociation constant between protein and free ligand, it is possible to calculate directly what the partition coefficient ought to be. One of the drawbacks of the affinity adsorbents is that although a fairly high degree of substitution can be achieved ($1-10$ $\mu$mol cm$^{-3}$), only a small proportion of the immobilized ligand seems to be oriented correctly even when spacer arms are used: typically only $1-2\%$ of potential sites actually bind the protein. Capacities of $1-2$ mg cm$^{-3}$ are regarded as good, which for a protein of MW 100,000 corresponds to an $m_t$ value of 0.01 to 0.02 mM. If we let $m_t$ be 0.02 and $p_t$ be 0.01 in Equation (4.5), then $\alpha$ values can be calculated for various $K_P$ values, as in Table 4.7.

The rather surprising result is that for good adsorption ($\alpha > 0.9$), $K_P$ has to be less than 0.001 mM, or $10^{-6}$ M, smaller than most protein-ligand dissociation constants. Considering that the specific interaction with an immobilized ligand on a column is likely to be weaker than with the free ligand, one may wonder how affinity adsorption chromatography ever works.

The answer to this dilemma is found in the extensive search for suitable methods of attaching the ligands to the supporting matrix. Direct attachment was rarely satisfactory, but use of a spacer arm, usually hexamethylene, gave good adsorption. A continuing problem was nonspecific hydrophobic interactions with the hexamethylene chain causing unwanted protein to adsorb to this spacer arm. Since the adsorption is hydrophobic in character, introduction of

**Table 4.7.** $\alpha$ Values Using $m_t = 0.02$ mM, $p_t = 0.01$ mM

| $K_P$ (mM) | $\alpha$ |
|---|---|
| 1.0 | 0.02 |
| 0.3 | 0.06 |
| 0.1 | 0.16 |
| 0.03 | 0.35 |
| 0.01 | 0.59 |
| 0.003 | 0.80 |
| 0.001 | 0.92 |
| 0.0003 | 0.97 |
| 0.0001 | 0.99 |

hydrophilic groups into the spacer arm (carbonyls, amides) should lessen the interactions and so make the adsorbent more specific for the protein for which it was designed. Unfortunately, these chemically highly sophisticated adsorbents often proved to have little attraction for the desired enzyme or anything else (83,84). As it turned out, the hydrophobic interactions were a necessary part of the binding to the adsorbent. Indeed, the original attempts at making affinity adsorbents without spacer arms may have failed not because the closeness of attachment of the ligand to the agarose backbone caused steric hindrance, but because some additional binding forces, provided by a hydrophobic spacer arm, were needed. Some thermodynamic calculations show the magnitude of additional forces that are required:

> The energy of interaction $\Delta G^0 = -RT \ln K_p$, and can be considered to be made up of the specific interaction between the enzyme and ligand, $\Delta G_L^0$, and nonspecific interactions $\Delta G_N^0$.
> If $K_p$ is 0.001 mM, then $\Delta G^0 = \Delta G_L^0 + \Delta G_N^0 = 34.5$ kJ mol$^{-1}$.
> If $K_L$, the specific dissociation constant is 0.1 mM, $\Delta G_L^0 = 23$ kJ mol$^{-1}$ and $\Delta G_N^0 = 11.5$ kJ mol$^{-1}$.
> Assuming $m_t = 0.02$ mM, and $p_t = 0.01$ mM, $\alpha$ can be calculated to be 0.92.
> Suppose that introduction of free ligand completely displaces all biospecific interaction with the matrix, so $\Delta G_L^0$ now equals zero. Then $\alpha$ goes from 0.92 to zero (since $\Delta G_N^0$ of 11.5 kJ mol$^{-1}$ is far too too weak to result in any significant retention).
> Putting in alternative parameters:

| $K_P$ (mM) | $K_L$ (mM) | $\alpha$ in absence of free ligand | $\alpha$ in presence of free ligand |
|---|---|---|---|
| 0.001 | 0.1 | 0.92 | 0 |
| 0.001 | 1.0 | 0.92 | 0.02 |
| 0.001 | 10 | 0.92 | 0.16 |
| 0.0001 | 0.1 | 0.99 | 0.02 |
| 0.0001 | 1.0 | 0.99 | 0.16 |
| 0.0001 | 10 | 0.99 | 0.59 |

It is seen that because of quite weak nonspecific forces, the $K_P$ value can be very small even if $K_L$ is of the order of $10^{-3}$ or higher. For example, AMP attached to agarose binds dehydrogenases tightly, even though the $K_i$ for AMP is usually in the range 1–10 mM for these enzymes; nonspecific interactions of about the same strength result in tight binding. As the forces are additive, the dissociation constants multiply together. Abolition or decrease of the specific binding by introduction of free ligand increases the $K_P$ value sufficiently for rapid elution. But if $K_L$ is high, the concentration of ligand needed will be higher still to displace the enzyme efficiently.

A convenient affinity adsorbent for hexokinases can be made by coupling glucosamine to agarose through an alkyl chain (85). N-aminoacyl glucosamines (in free solution) were found to have $K_i$ values greater than for glucosamine itself when the number of C atoms in the aminoacyl chain was 6 or less, indicating disturbance of binding caused by the acyl chain, but when 8 C atoms were present the $K_i$ value was much lower, i.e., some enzyme–acyl chain inter-action was occurring. Whereas yeast hexokinase would not bind to agarose-$(CH_2)_6$-acylglucosamine, it was retarded on a column of agarose-$(CH_2)_8$-acyl-glucosamine and could be eluted by glucose. The nonspecific interaction was needed for binding to the adsorbent. Mammalian hexokinase isoenzymes showed selective binding to these adsorbents; types II and III could be chro-matographed on adsorbents with $C_6$ spacer arms, whereas the kidney type I isoenzyme needed $C_8$ for retardation.

The calculations made in Tables 4.7 and above are based on two assump-tions that may not be justified. The first concerns the possible heterogeneity of distribution of attached ligands; the $m_t$ value may be much higher in restricted portions of the absorbent, so that much more protein can bind even with weak binding constants (e.g., $K_P \sim 10^{-4}$ M). A recent investigation of this possibility utilized spin-labeled model ligands to determine distribution (86). It was found that inhomogeneity was not extensive. However, it was revealed (as can be calculated from substitution levels) that the closeness of the attached ligands permitted multiple attachment to an average-sized protein, provided that the protein had more than one binding site. This leads to the second assumption above—single-site attachment may not be the whole story, and the low occu-pancy may in some cases be due to the requirement for just the correct spacing between adjacent ligands to allow two-site attachment (Figure 4.40). This could not of course apply to monovalent (single active site) enzymes.

In conclusion, it can be seen that affinity adsorption chromatography needs a total energy of interaction between protein and matrix of at least 35 kJ mol$^{-1}$, which can rarely be supplied by a single protein–ligand interaction. Nonspe-cific interactions may amount to half or more of this value, due to mainly hydrophobic forces between surface residues on the protein and the spacer arm. In this case affinity elution should be effective, and quite weak specific inter-actions are adequate. The low-percentage occupancy of affinity adsorbents may

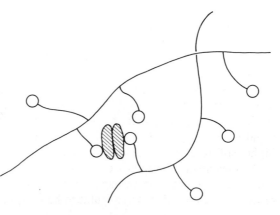

**Figure 4.40.** Exact spatial requirement for two-point specific attachment to di-meric enzyme on an affinity adsorbent (enzyme shaded).

in some cases be explained by the requirement for two-point biospecific attachment (Figure 4.40), but is more likely to be due to a requirement for nonspecific in addition to biospecific binding. If these two attachments have rigorous spatial requirements, relatively few ligands will be correctly oriented to provide a protein binding site (Figure 4.41).

In view of the requirement for nonspecific interactions, it is interesting to speculate on the possibility of purposely introducing them, by either using a matrix which is more hydrophobic or including a high density of other attracting groups, e.g., DEAE (Figure 4.42). The latter type of adsorbent would be operated in conditions where the enzyme would be hardly bound at all on a simple DEAE–ion exchange column ($\alpha$ about 0.1–0.4), but the ion exchange

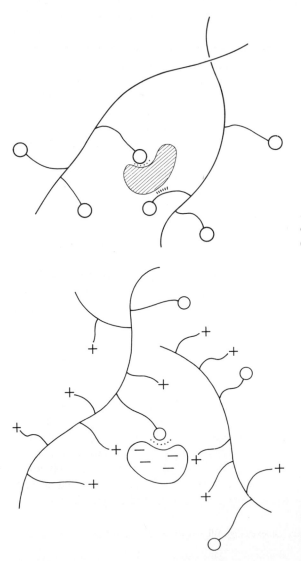

**Figure 4.41.** Exact spatial requirement for both specific and nonspecific attachment on an affinity adsorbent (enzyme shaded).

**Figure 4.42.** Combined affinity and ion exchange adsorption for negatively charged proteins.

interaction would be a sufficient addition to the specific interaction to make $\alpha$ 0.99 or higher on the combined adsorbent. Affinity elution would then work well, and the capacity for the enzyme in question would be high. The procedure is as follows:

(1) Adsorb and elute from DEAE-cellulose by conventional ion exchange chromatography.
(2) Apply the eluted fraction without buffer change to the affinity-DEAE adsorbent—only the enzyme specifically required would bind.
(3) Elute with ligand in buffer, or by increasing ionic strength, lowering pH.

Such hypothetical mixed adsorbents have not been made,* so this idea remains a conjecture at present. As many ligands are negatively charged, cation exchange effects may automatically be operating, and although it is generally advised to operate the system at ionic strengths and pH values to avoid these effects, being idealistic may not be the right approach! For instance, a kinase which has a weak affinity for AMP may not bind to AMP-agarose in 0.1 M phosphate buffer, but may, with the combination of cation exchange due to negative charges on the AMP, bind in a weaker buffer such as 0.01 M Mes or Mops. Although other basic enzymes may bind also, there is no reason why they should be affinity-eluted by introduction of free ATP used to elute the kinase, provided that the concentration to ATP used does not significantly raise the ionic strength.

## General Techniques and Procedures in Affinity Adsorption Chromatography

In the early 1970s when affinity chromatography was proving highly successful in many instances, the feelings among enzyme chemists were, "Why did we not think of it before?" and "This will make all other methods obsolete." The general euphoria dissipated later after many people had spent months, often years, trying to develop a suitable adsorbent for their particular enzyme and often not succeeding. For some applications, especially when the enzyme has a unique substrate, and it makes up only a fraction of the total protein present, once a suitable adsorbent has been found, there could never be any question of not using the technique. One-step purifications of 1000-fold with nearly 100% recovery have been reported; the ease and speed of the method has revolutionized, indeed made possible, the isolation of a large number of important enzymes occurring in small amounts. But for every success story there is at least one complete failure and a lot of wasted effort. The main problems have been the low capacity of the adsorbents, though newer methods of attaching ligands are improving this factor, and nonspecific adsorption (which may lower the capacity for the specific enzyme still more). However, as we have seen, nonspecific interactions are a necessary evil if the desired enzyme is to adsorb,

---

*Bio-Rad have recently introduced ion exchange Blue-agarose adsorbents based on this principle.

so very careful trial-and error designs of adsorbent and buffer conditions are needed to reach the ideal conditions.

To design an adsorbent we must first consider the ligands and their likely mode of interaction with the protein. If possible the attachment should be at a point that is not involved in the protein/ligand interaction; some knowledge of substrate analogues and their $K_i$ values can help here. If there is a choice of two ligands, a large one and a small one, choose the large one, as there is more opportunity for varying the mode of attachment to the matrix. A small molecule such as alanine is unlikely to be attached successfully, since an enzyme recognizing alanine would almost certainly make use of both its charged groups and would recognize the methyl as being distinct from other amino acids. Thus the chances of the methods of attachment illustrated in Figure 4.43 *not* destroying the enzyme's ability to recognize the molecule are slim. On the other hand, adenine nucleotides have been successfully attached to agarose through a number of positions (Figure 4.44).

The starting material can be plain agarose, or preactivated (CNBr, epoxy, carboxyldiimidazole, tosyl), or with a hexyl side-chain terminating in carboxyl or amino. Reactions with excess of ligand or analogue may take place in organic or aqueous solvents depending on the chemistry involved. After carrying out the chemistry, an estimation should be made to determine the quantity of ligand (if any) that has been immobilized.

Trials of the effectiveness of the adsorbent are usually carried out in a Pasteur pipette; a small sample (not more than 1 ml) of enzyme in suitable buffer is applied, and the column washed. If the enzyme sticks under these conditions, one can assume that $\alpha$ is greater than 0.9 and the adsorption has been achieved rapidly, since a Pasteur pipette packed with affinity adsorbent has a flow rate rather higher than optimal. Trials for elution can now be carried out. The buffer used initially may be important; as indicated above, *avoidance* of ion exchange effects may not be the best practice (unless it turns out that there is a very strong biospecific affinity), and a range of different ionic strengths, buffers, and salts may be tried. For elution there are three main methods. Affinity

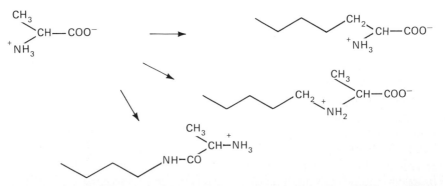

**Figure 4.43.** Possible ways of immobilizing alanine as a ligand for an affinity adsorbent.

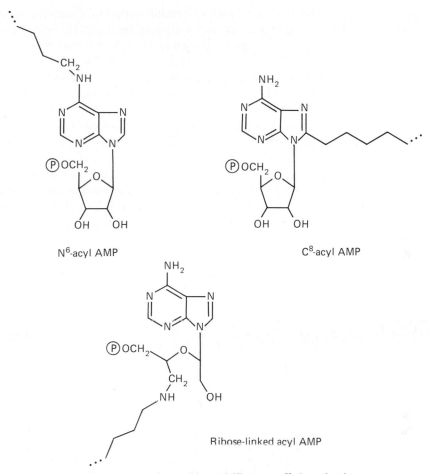

$N^6$-acyl AMP          $C^8$-acyl AMP

Ribose-linked acyl AMP

**Figure 4.44.** Three ways of attaching AMP as an affinity adsorbent.

elution, by inclusion of free ligand in the buffer, is the ideal method, but can sometimes prove expensive. Often quite high concentrations of ligand have been used; this has been because the interaction with the column was so strong ($\alpha > 0.999$) in these particular buffer conditions that a large value of $\ell$ (see Eq. 4.9) was needed. If the buffer is first altered so that $\alpha$ is reduced to around 0.9, much lower concentration of ligand can be used, as described for affinity elution from ion exchangers.

To achieve a lowering of $\alpha$ before affinity elution, some knowledge of the interactions involved is needed. Increasing salt concentration may well decrease the biospecific component of binding, but it may also decrease the subsequent interaction with free ligand in affinity elution (increase the value of $K_L$). Increased salt would minimize any nonspecific ionic interactions, but it would also increase hydrophobic interactions, especially with the spacer arm. If the latter are very important, *decreasing* salt concentration may be more effective.

Alternatively, introduction of a surface tension–reducing agent (to lessen hydrophobic and van der Waals interactions) may be used. This could be a nonionic detergent such as Triton X-100, up to 1% v/v, ethylene glycol (10–50% v/v), or a relatively small amount of ethanol (up to 5% v/v). The latter is unlikely to affect the value of $K_L$ with charged ligands. After a prewash with this buffer, ligand is introduced in the same buffer for the affinity elution process.

A second method is to use a high salt concentration, which disrupts all non-hydrophobic interactions between immobilized ligand and protein, including any biospecific ionic or other polar forces. Increasing salt concentration increases hydrophobic forces, so one disadvantage is that the nonspecific interactions may become stronger. Nevertheless, salt usually displaces the protein. In some cases very tight binding occurs and neither free ligand nor salt will dislodge the protein (this is particularly true for immunoadsorbents; cf. section 4.7). Chaotropic salts such as thiocyanate at high concentration can be successful in these cases; alternatively, a low pH (but now low enough to denature the protein) may be best.

Other methods for elution include temperature changes (decreasing temperature decreases hydrophobic forces), dilution, even washing with water (cf. section 4.7), and electrophoresis to remove charged proteins from the column (87). In conclusion, affinity adsorption chromatography is an excellent method once the procedure has been worked out, but starting from scratch, development of a suitable adsorbent may be a very slow process.

Many ready-made adsorbents are available commercially, and are suitable for a large range of proteins. Because of the low capacity and expense of the adsorbents, the method is most useful for purifying proteins that make up only a small proportion of the total protein present, in which case it can be used as the first (and sometimes only) step in purification. On the other hand, for dealing with large quantities of the required enzyme (e.g., > 100 mg), other methods may be more suitable, at least as preliminary steps in the fractionation procedure, with an affinity adsorption procedure placed only at the end for a final improvement of the preparation. The adsorbents can be re-used many times provided that they are cleaned immediately after use (e.g., urea or sodium dodecyl sulfate solutions), and stored in the presence of bacteriocidal agents (azide, merthiolate, chlorbutol). However, leakage of ligand due to the inherent instability of the isourea linkage formed from cyanogen bromide activation has been a problem. This can be overcome by using one of the other activation processes; each results in a more stable linkage.

In view of the many excellent reviews and monographs on affinity chromatography (see end of chapter for a selection), I have not attempted a more detailed coverage of this topic, nor to list the many different sorts of adsorbent that have been developed. However, the particular example of immunoadsorbents, in which the "ligand" is an antibody to the protein being purified, is dealt with separately in section 4.7.

## 4.6  Dye Ligand Chromatography

Pharmacia Fine Chemicals attached a blue dye known as Cibacron Blue F3GA
to a high-molecular-weight soluble dextran for use as a visible void volume
marker for gel filtration. Had they selected a different dye, it is possible that
the recent rapid expansion in the use of colored adsorbents would not yet have
happened. Workers were puzzled when yeast pyruvate kinase appeared in the
void volume during gel filtration, indicating a much higher molecular weight
than expected. But when the blue marker was omitted, the enzyme came out
after the void volume (88). By carrying out gel filtration first in the absence of
marker, and then in its presence, purification was achieved—although the
marker had to be removed afterwards. The pyruvate kinase was binding to the
blue dye; by coupling the blue dextran to agarose by standard affinity chromatographic methods, a new type of adsorbent was produced. Similar observations were made almost simultaneously concerning yeast phosphofructokinase (89). Soon afterwards other enzymes, especially dehydrogenases, were
found to have an affinity for this blue dye, and it is now generally accepted that
most enzymes that bind a purine nucleotide will show an affinity for Cibacron
Blue F3GA. Although many other related dyes also bind to various enzymes,
the original blue one has been used in the vast majority of cases to date.

The specificity of dye adsorbents is such that they can qualify as affinity
adsorbents, even though the dye used bears little obvious resemblance to the
true ligand; "pseudo-affinity adsorbents" is perhaps a better description.
Molecular models do show a rough resemblance between Blue F3GA and
NAD (Figure 4.45), but the most important correspondences are with the
planar ring structure and the negative charged groups. It has recently been
shown by X-ray crystallography that Blue F3GA binds to liver alcohol dehy-

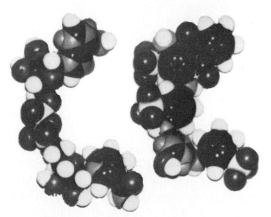

**Figure 4.45.** Three-dimensional structural similarities between Cibacron Blue F3GA
and NAD. On the left is a CPK model of NAD, and on the right Cibacron Blue F3GA
oriented in a similar fashion. [From Thompson et al. (90).]

drogenase in an NAD site, with correspondences of the adenine and ribose rings, but not the nicotinamide (91). Thus the dye behaves as an analogue of ADP-ribose, and it binds to AMP, IMP, ATP, NAD, NADP, and GTP sites. The binding is commonly tighter than the true substrate, with $K_i$ values in the micro-molar range. As a result, nonspecific interactions are not always needed to enable an enzyme to bind to an immobilized dye column. Moreover, although the original adsorbent used was "blue dextran," the dextran component only acts as something to couple to the agarose, a sort of giant globular spacer arm—in fact, so large that much of the inside of the agarose beads would not be accessible by it. Even so, the binding capacity of blue dextran agarose is quite high, with a much higher percentage of the attached dye ligand being active in binding proteins than is the case with a normal affinity adsorbent.

Coupling blue dextran to agarose is inconveniently complex. The dye itself is sufficiently reactive to couple directly to nonactivated agarose. Indeed, this dye and the others described below were developed for the purpose of reactive coupling to biological polymers (wool, cotton), and do so through the triazinyl chloride group, in slightly alkaline conditions (Figure 4.46).

Agarose provides a great number of hydroxyls for the reaction, and direct coupling can be manipulated to provide a range of substitution percentages,

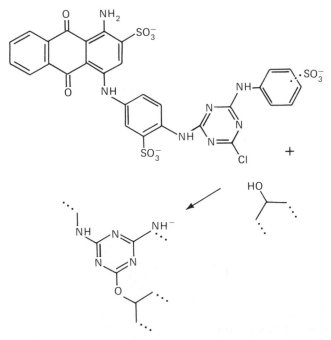

**Figure 4.46.** The structure of Cibacron Blue F3GA and its reaction with hydroxyl groups.

usually up to about 5 $\mu$mol cm$^{-3}$ (swollen gel). Capacity for proteins is some 10–20 times that for comparable true affinity adsorbent; 20–30 mg cm$^{-3}$ of low-molecular-weight proteins will bind, but due to gel exclusion effects the capacity decreases as the protein size increases. At low ionic strength and in a lower pH range the dye adsorbents are also very effective cation exchangers, as the multiple negative charges cause binding of basic proteins that do not necessarily have a specific nucleotide binding site. Unless there is a reason for making use of this property, the dye ligand adsorbents should be used in conditions where cation exchange behavior is weak, e.g., pH 7 or more, and an ionic strength of at least 0.1.

Note that the direct attachment to the matrix backbone does not detract from the dye's ability to bind enzymes: a spacer arm does not seem to be necessary. As was mentioned in the previous section, the reason why spacer arms are needed when weak enzyme–ligand interactions are involved may be not so much to position the ligand further away from the matrix as to provide nonspecific interaction forces to strengthen the binding. With typical $K_i$ ($K_P$) around $10^{-6}$ M, and a higher value for $m_t$, it is seen (from Eq. 4.5) that strong adsorption does not need extra interactions for dye ligand columns—e.g.,

$$m_t = 0.1 \text{ mM}, \qquad p_t = 0.05 \text{ mM}, \qquad K_P = 0.001 \text{ mM}, \qquad \alpha = 0.98.$$

From this simple description it would seem that dye adsorbents should be ideal for nucleotide-binding proteins, especially in combination with affinity elution. In practice there are problems, the main one being nonspecific binding forces which can make affinity elution less efficient than it should be. In addition, there are so many nucleotide-binding proteins that adsorbing them *all* would not give a high degree of purification.

During recent years the investigation of a wider variety of dyes has led to improvements in selectivity, since the affinities for different dyes vary considerably depending on the enzyme concerned. The first noted selectivity was that NADP-dependent enzymes bind more strongly to Procion Red HE-3B than to Cibacron Blue F3GA, whereas with most NAD-dependent enzymes the converse was true (92). Later, Amicon Corporation introduced three more adsorbents from a range they tested (93). Other tests with ranges of dyes for specific enzymes have been carried out (94–96). It is perhaps unfortunate that there is now such a range of possibilities (including operational parameters, see below), that dye ligand chromatography becomes quite a complicated operation compared with the early days of "see if it sticks to blue dextran." Amicon has introduced a screening kit of 5 dyes in their Mātrex range; up to 90 different dyes have been tested for adsorption of a range of enzymes (97).

There are some spectacular examples of successful application of dye adsorbents, but in most cases it is just used as a step in purification, preceded and followed by others. Proper application of the principles of affinity elution chromatography would undoubtedly improve some of the steps that have been reported.

## Preparation of Dye Ligand Adsorbent

Three factors describe a dye ligand adsorbent: (1) the nature of the matrix, (2) the structure of the dye, and (3) the degree of substitution of dye. Taking these in turn, firstly consider the matrix. Like affinity adsorbents, it must have an open porous structure which allows large proteins to penetrate the particles; it should have a good flow rate when packed in a column, and it should preferably be inert when unsubstituted. Although a number of other matrices have been used for specialized purposes, agarose gels are most widely used and do not have any disadvantageous properties. Dextran gels (Sephadex) have been used, but in order to obtain a good penetration by proteins they must be very open gels such as Sephadex G-150 and G-200, which are soft and have poor flow rate characteristics. Agaroses are more porous. However, since the coupling of the dye proceeds best at temperatures above 40°C, and agarose melts above this temperature, the adsorbents based on plain agarose beads may have relatively low degrees of substitution. Cross-linked agaroses, either with epichlorhydrin, 2,3-dibromopropanol, or acrylamide chains, are the most suitable. Such commercial products as Sepharose CL-4B, CL-6B, Sephacryl S-300, Ultrogel, and Bio-Gel A are all suitable materials for dye ligand adsorbents; several of them are unharmed at 100°C.

The dyes used are the reactive triazine dyes known by a variety of names; the structures are not always known precisely because of patent rights. The ICI series is known as Procion, whereas Ciba-Geigy supplies the overlapping Cibacron series. The dyes also have color index numbers and are known as "reactive (color) (number)." Thus the most widely used "Cibacron Blue F3GA" is also known as "Procion Blue H-B," "Color Index 61211," and "reactive blue 2" (structure in Figure 4.46). The next most widely used dye is Procion Red HE-3B (reactive red 120) which has shown more specificity toward

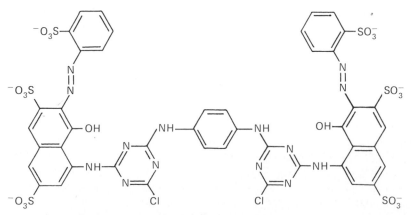

**Figure 4.47.** The structure of Procion Red HE-3B (reactive red 120). This is a bifunctional dye; one of the triazinyl chlorides attaches to the agarose when synthesizing the adsorbent, whereas the other may be hydrolysed.

NADP-binding enzymes (Figure 4.47). Brown, green, orange, yellow, and other blue and red dyes have been used for particular purposes. With such a potential variety, the possibilities of selectivity for each particular enzyme mixture become very numerous. Nevertheless, the basic principles of adsorption through hydrophobic and ionic attraction, principally with nucleotide-binding sites, applies to all of these dyes. At present, Cibacron Blue F3GA and Procion Red HE-3B coupled to cross-linked agarose with a high degree of substitution are available commercially from several sources. In addition, "blue dextran agarose" is available. The Mātrex range includes the above two dyes, together with a phthalocyanin-type blue, a green, and an orange.

To couple the dyes directly to agarose, the gel is treated with the dye solution at pH 10-11 ($Na_2CO_3$ or NaOH) in the presence of salt (2% NaCl). At 30-40°C, coupling is complete in 2-3 days for the monochlorotriazine dyes (ICI Procion H series), but dichlorotriazine dyes (ICI Procion MX series) are more reactive, and coupling is completed in 1-2 hr. Most excess dye hydrolyses to an inactive form (97). Using cross-linked agarose, the solution can be heated to higher temperatures. It has been found that 2 hr at 80-90°C using a 0.2% dye solution, followed by a further fresh 0.2 g of dye per 100 ml and heating for another 2 hr, gives a very high degree of substitution using Procion H dyes. The mixture is allowed to cool overnight, and excess dye washed off using a Büchner funnel with large amounts of water. The dye adsorbent may be stored in a salt solution with a bacteriostatic agent such as azide; before and after use it should be treated with a solution of 6 M urea—0.5 M NaOH, which removes any loosely bound dye remaining.

The degree of substitution is an important feature in dye ligand chromatography, as it is in other affinity methods. If more than one dye substituent can be involved in the binding, then much tighter adsorption is obtained (see Figure 4.40). Moreover, the higher the value of $m_t$, the higher is $\alpha$ (Eq. 4.5). It may be that a highly substituted gel binds the enzyme required so tightly that subsequent recovery is low, in which case more weakly substituted gels can be made by shortening the treatment with dye during their preparation. For consistent results the degree of substitution should be recorded (98).

## Operation of Dye Ligand Adsorbent Column

As with any column chromatography, the aim is to adsorb the protein of interest, then elute it by a controlled scheme minimizing contamination by other proteins. Since so many proteins can potentially bind to dye adsorbents, starting conditions should be found that minimize the adsorption of other proteins while retaining all of the desired one on the column. Adsorption generally weakens with increasing ionic strength; the ion exchange behavior seems more important than the hydrophobic attraction for proteins. As a result, pH is also more important than with normal affinity adsorption columns, and less strong adsorption occurs at higher pH. With a combination of both specific and nonspecific binding forces, enzymes can be so strongly bound that they are difficult to elute. For instance, if the hydrophobic contribution is large, increasing salt concentration will strengthen it and counter the displacing effect salt has on the ion exchange and specific adsorption behavior. Inclusion of solutes to counter hydrophobic forces such as detergents (Triton X-100), chaotropic salts (LiBr, KCNS), or glycols may be useful in eluting such proteins. A final wash with water also elutes some proteins (Figure 4.48).

For affinity elution, the true substrate should be used; although free dye should work in theory, it would be less specific. Thus for kinases, ATP is used ($\pm$ Mg), for dehydrogenases $NAD^+$ (or preferably NADH, which binds more tightly to most dehydrogenases), AMP-binding enzymes (e.g., fructose 1,6-bis-phosphatase), AMP. But before wasting high concentrations of expensive ligands, the best conditions for affinity elution should be found, e.g., in a buffer where the enzyme is just beginning to move down the column, $\alpha$ between 0.95 and 0.90. Addition of ligand at a concentration of no more than $10 \times K_L$ if the enzyme has only one binding site, or less if it has multiple binding sites (see Figure 4.28 and 4.29), should cause rapid elution. There are many reported examples where ATP or NAD has been used at 10 mM or even more to displace the enzyme, simply because the elution buffer composition was not ideal; such methods would be completely uneconomical on a large scale.

It has been proposed that more efficient operation can be achieved using a two-column method (96); the first contains a dye which does not cause binding of the enzyme required. This column removes very strongly attracted proteins which would otherwise adsorb to and occupy potential sites on the second column. The second column should contain a dye that has a strong affinity for the enzyme required, but relatively low for others. To find the optimum conditions requires an extensive screening with many dye ligand columns, measuring both enzyme and protein before and after passage through the column, and after

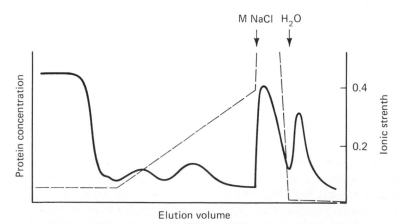

**Figure 4.48.** Elution of proteins from a Cibacron Blue F3GA-agarose column. Sample applied: crude extract of *Zymomonas mobilis,* in pH 6.5 buffer $I = 0.08$. Salt gradient (dashed line) to 0.4 M NaCl applied, followed by M NaCl, then washed with water. (From author's unpublished observation.)

elution with some appropriate buffer. This procedure should be considered in cases where otherwise a very large amount of adsorbent appears to be required for the enzyme in question.

Finally it should be noted that there are examples of proteins which do not bind functional nucleotides, yet adsorb to dye columns. Serum albumin is the most notable, but others such as follicle-stimulating hormone, interferon, serum lipoproteins, phosphoglycerate mutase, and fructose 1,6-bisphosphate aldolase bind also (93). In the latter two cases, the enzyme substrates (2,3-bisphosphoglycerate and fructose 1,6-bisphosphate, respectively) can be used for affinity elution. Other than nucleotide-binding proteins, enzymes with polyanionic substrates are most likely to bind and to be eluted by their substrates.

The capacity of the columns may be quite large, but the amount of the desired enzyme that binds will depend considerably on the amounts of other enzymes that are present in the mixture. Thus, at an early stage in purification very large quantities of adsorbent might be needed, whereas at a late stage a small column would suffice. Crude extract can be used only for the most tightly binding proteins, since they will displace any weaker ones. Note that, if crude extracts contain phenols or flavins, some dye adsorbents bind these also, and very extensive washing may be needed to remove all UV-absorbing material. Once again, trial and error is the procedure to be adopted. Adsorption methods are not instantaneous, and dye adsorbents are no exception to this rule. Sufficient time should be allowed for the solution to equilibrate with the matrix, and flow rates in columns should normally be restricted to not more than 20 cm hr$^{-1}$. On the other hand, a very strongly adsorbing enzyme can be purified with fast flow rates. Perhaps the most extreme example of this reported is for yeast AMP kinase (12), where a flow rate of 40 ml *min*$^{-1}$ was used with a column of "Affi-Blue" only 2.5 cm in diameter—nearly 500 cm *hr*$^{-1}$! Similar flow rates have been used for pumping crude yeast extracts through a column of Procion HE-3B/Sepharose CL-4B, which retains all the glucose 6-P dehydrogenase and most of the 6-P gluconate dehydrogenase. For such strongly adsorbing enzymes, batch-wise techniques (cf. section 3.9) may be more appropriate, or else use of dyes such as Procion Yellow H-A (Amicon Mātrex orange) or Procion Scarlet MX-G, which do not bind so much protein as most other dye ligand adsorbents.

Dye ligand chromatography is a fast-growing field, and during the preparation of this chapter new publications together with work in my own laboratory have convinced me that the method should be considered at the outset of an enzyme purification project—especially so if the enzyme has a nucleotide ligand. There appears to be nothing special about Cibacron Blue compared with other dyes; it falls somewhere in the middle of the range as far as ability to bind proteins is concerned. Two important factors that make these adsorbents so useful are (a) their relatively high binding capacity and (b) their cheapness to produce. If these are important considerations for the problem at hand, then dye ligands should certainly be investigated before conventional affinity adsorbents.

# 4.7  Immunoadsorbents

The ultimate specific adsorbent for any enzyme or protein is an antibody raised against that protein, attached to an insoluble matrix. Antibodies are highly specific and have very high affinities for the proteins they have been raised against. Consequently a successful antibody column has features that are generally more selective than other sorts of affinity adsorbents; immunoadsorbents are specialized types of the biospecific adsorbents described in section 4.5.

Because the raising of antibodies is a complex process, immunoadsorbent columns have not been widely used yet. But the new technology of monoclonal antibody production may revolutionize this aspect of enzyme purification. The following first describes the techniques and problems associated with conventional, polyclonal antibodies, then briefly outlines the ways in which most of these problems can be overcome.

## Basic Principles

These are outlined thus:

1. Purify enzyme by conventional techniques, from a source evolutionarily distant from the immunization animal, usually a rabbit.
2. Raise antibody to enzyme in rabbit using conventional techniques; bleed rabbits and test for antibody by immunoprecipitation, double diffusion, or a far more sensitive radiochemical method involving detection of antibody–antigen aggregates using $^{125}$I-labeled "protein A"—protein A is a protein from *Staphylococcus aureus* which binds tightly to immunoglobulin IgG chains (99).
3. Partially purify antibody from rabbit serum (precipitation at 33% saturation of ammonium sulfate is normally sufficient).
4. Couple antibody to adsorbent, e.g., cyanogen–bromide-activated agarose, tosyl-activated agarose (cf. section 4.5).
5. Run crude sample, or partially purified sample containing enzyme through antibody column. Wash column.
6. Elute column.

The difficult steps are usually 1, 2, and 6. We will assume that step 1 has been achieved, at least to a partial purification.

Less than a milligram of pure enzyme is sufficient to raise good antibodies in a rabbit. Maintenance of a high antibody titer in the blood can be achieved by additional monthly injections of 0.1–0.3 mg enzyme. However, some proteins are far more antigenic than others, so much so that even traces of an impurity in the enzyme preparation, undetectable by normal analytical techniques (cf. section 9.1), may also raise quite strong antibodies in the rabbit. Subsequently these antibodies would then combine with their antigen, which in a crude preparation would be present in larger amounts. Hence, the selectivity of the antibody preparation can be far poorer than expected. Such problems are particularly acute when the desired enzyme is a relatively poor antigen.

One quite successful technique which can overcome this problem is to carry out a gel-electrophoretic separation of the desired enzyme from other proteins (100). The enzyme sample is run into a fairly large-cross-section gel and electrophoresis continued until the enzyme is clearly separated from other impurities. After locating the enzyme, the gel is sliced and the section containing the enzyme is mashed to give a fine suspension of polyacrylamide gel plus enzyme. This is then used for the first injection; a few hundred micrograms of protein is sufficient.

Immunoglobulins raised in this way can consist of as much as 10% of antibodies to the enzyme required, the remaining 90% being relatively nonspecific immunoglobulins that happened to be present in the animal at the time. Tens of milligrams of immunoglobulin can be attached per $cm^3$ of agarose (101); up to 1 mg of specific antibody per $cm^3$ may be present. However, since the attachment of the antibody molecules to the adsorbent will in most cases be in a way that prevents it binding to its antigen, perhaps only $\pm 0.1$ mg $cm^{-3}$ is effective. If one of these antibodies binds one molecule of antigen of a similar molecular weight, then the capacity of such a column would be of the order of 0.1 mg $cm^{-3}$— not great, but nevertheless very useful for purifying small amounts of enzyme from a crude mixture. One rabbit may provide as much as 1–2 g of immunoglobulins: this might make several hundred $cm^3$ of adsorbent, sufficient to purify at least 10 mg of enzyme *each time it is used*. Thus the potential problem of not regaining the amount of enzyme used in making the antibody is not serious, especially if the column can be used more than once.

The final step, elution from the column, has often proved to be the stumbling block in the whole process. Antigen–antibody interactions characteristically have dissociation constants of the order of $10^{-8}$–$10^{-12}$ M. Putting this value as $K_P$ in Eq. (4.5) gives $\alpha$ values of effectively unity. To reduce the specific interaction is not easy. Although immunoglobulins are very hardy molecules and can stand many conditions extreme enough to disrupt the specific interaction, that cannot necessarily be said of the enzyme being displaced. Thus a pH of 2–3 will displace the antigen, leaving active antibody on the column capable of being re-used, but more likely than not pH 3 will destroy the enzyme. A similar fate is usually seen with high concentrations of urea or organic solvents to minimize H-bonding and hydrophobic forces, respectively. High salt concentration may be successful if the interaction happens to be largely electrostatic, but conversely it will strengthen any hydrophobic bonds. Chaotropic salts such as thiocyanate or iodide, lithium bromide, magnesium chloride and so-called deforming buffers such as imidazole citrate have been used with some success, but again the concentration needed can cause denaturation of the enzyme once displaced. A rapid elution procedure for obtaining active antigen after elution with a denaturing agent has been reported (102). In this technique the adsorbent is placed on top of a desalting column, so that as soon as the enzyme emerges from the adsorbent it separates from the eluting agent (Figure 4.49). On a small scale exposure to the eluting agent may be no more than 1 min, in which time little denaturation may be expected.

Recently some reports of successful elutions using *cold distilled water* have been reported (103,104). In both cases the enzymes concerned were plant enzymes. How general this method might prove remains to be seen, since there is no obvious reason why it should have worked, except for considerably weakening hydrophobic forces; it may have been a specific conformational response of the *enzyme* to the low ionic strength.

Another technique for displacing the antigen is electrophoresis (87). At a pH where the antigen is sufficiently charged, electrophoretic mobilization of the small fraction of antigen in solution slowly moves it away from the adsorbent without the excessive dilution that simply passing buffer through would cause. A novel technique involves the use of a cleavable spacer arm (105). The antibody is attached through a phenyl ester linkage to the agarose matrix. This is chemically cleaved with an imidazole-glycine buffer at pH 7.4 to release the antibody–antigen complex. Unfortunately this does not separate the antibody from antigens, and the adsorbent can only be used once.

Ideally, removal of adsorbed enzymes from columns would occur by weakening of the dissociation constant from a value of about $10^{-7}$ M to around $10^{-4}$ M, about a 1000-fold decrease in affinity, without causing denaturation of the antigen. In order to start with a value around $10^{-7}$ M (rather than the more usual affinity of several orders of magnitude stronger), it may be practical to vary the species from which the original enzyme sample was purified. For example, if a human enzyme is the required product, but the human enzyme–

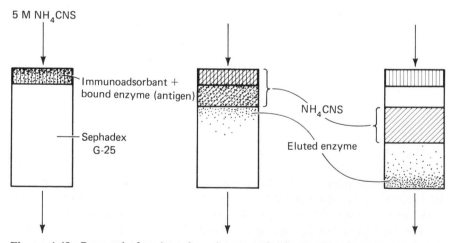

**Figure 4.49.** Removal of antigen from immunoadsorbent using chaotropic salt and rapid separation of salt from antigen (protein being purified).

rabbit antibody complex is too strong, the initial enzyme might be purified from another animal, perhaps a horse. The horse enzyme–rabbit antihorse antibody complex would be strong, but the human enzyme would not cross-react with the antihorse antibody so well. Consequently purifying the human enzyme with the antihorse antibody attached to the column may be more successful. It would also have the practical advantage that a large amount of more convenient raw material (i.e., a horse) can be used for the initial enzyme purification processes.

## Monoclonal Antibodies

Most of the problems encountered using polyclonal antibodies can be avoided with the new technology of monoclonal antibody production (106). The facilities needed and the techniques used, although simple enough, are more demanding than those required for ordinary antibody production. The advantages over polyclonal antibodies for making immunoadsorbents are:

(1) The enzyme preparation used does not have to be pure; the screening process for detecting suitable clones rejects antibodies directed against other proteins. Here 1% "purity" can be sufficient (107).
(2) As mice are usually used, the amount of antigen required is very small; a total of 50 $\mu$g may be sufficient.
(3) Many different clones of antibody types against the desired enzyme may be found, directed against different antigenic determinants. The dissociation constants for the antibody–enzyme complex will vary, and it is possible to select a clone giving $K_d$ in the desirable range of $10^{-6}$–$10^{-8}$ M, so avoiding problems when eluting columns.
(4) Once a suitable clone has been found, storage in liquid nitrogen enables an indefinite supply of that antibody to be obtained from many individual animals; the supply does not terminate with one animal's death.
(5) As the monoclonal antibody is pure, *all* the coupled immunoglobulin on the adsorbent is specific for the desired enzyme, so capacities of the immunoadsorbent are generally at least 10-fold higher.

The first, spectacular successes in using monoclonal immunoadsorbents were in the purification of human interferon (107) and of the sucrase/isomaltase of intestinal brush borders (108). By the time this appears in print there will no doubt be more examples reported.

Because of their high affinity for the desired enzyme, immunoadsorbents (polyclonal and monoclonal) are well suited to use in batch-wise adsorption techniques (cf. section 4.9). But the formation of a complex between antigen and antibody is usually a slow process, so whether working batch-wise or in a column, considerable time should be allowed for adsorption to occur. During the elution process also very slow flow rates should be employed; even stopping of the flow altogether may be useful (104).

# 4.8  Miscellaneous Adsorbents

## Inorganic Adsorbents

All adsorbents discussed so far in this chapter have been based on biological matrices, with occasional reference to synthetic organic polymers. There is also a range of inorganic substances that have been used for protein adsorption, mainly oxides, insoluble hydroxides, and phosphates. Chief among these is calcium hydroxyphosphate, which in crystalline form is known as hydroxyapatite. The use of hydroxyapatite will be described below, as will the use of the gelatinous form "calcium phosphate gel" in the final section of this chapter under batch adsorption. Meanwhile, a short listing of adsorbents will give an idea of the usefulness of inorganic materials. One particular advantage they have, especially in large-scale and industrial applications, is their cheapness; it is often not worth the trouble to clean them up after one use.

Unlike ion exchangers or affinity adsorbents, the inorganic materials do not have a readily explainable mode of action. Crystalline surfaces are made up of charged ions, with associated water of hydration, so undoubtedly an electrostatic interaction is an important component of the adsorption of proteins to them. However, every positive area on the crystal surface has negative areas very close to it, and vice versa: consequently the interaction is more likely to be a polar dipole–dipole bonding than an ion exchange effect (Figure 4.50a). Proteins have the greatest probability of having adjacent positive and negative groups in the neutral pH range; thus, whatever the isoelectric point may be, adsorption is most likely between pH 6 and 9.

Despite this simple interpretation, the practice is more complex. Buffers are always present, usually phosphate, and the buffer ions themselves adsorb to the inorganic materials. Thus it is possible for hydroxyapatite to present an entirely negatively charged surface to a protein (Figure 4.50b). Increasing buffer concentration causes competition between protein and buffer ions for the charged sites on the adsorbent, resembling typical ion exchange behavior. Detailed discussions of these effects have been presented (109,110) with some quantitative data on the adsorption of certain proteins to both hydroxyapatite and $TiO_2$.

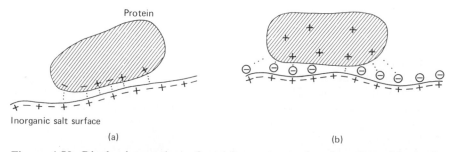

**Figure 4.50.** Dipolar interactions of proteins on inorganic salt surface: (a) simple interaction, (b) with intervention of buffer ion.

**Table 4.8.** Some Inorganic Adsorbents That Have Frequently Been Used for Protein Purification

Alumina gel $C_\gamma$ (gel and crystalline)
Bentonite (a silicaceous powder)
Calcium phosphate:    Aged gel
                      Brushite
                      Hydroxyapatite
Titanium oxide
Zinc hydroxide gel

As with many other enzyme purification procedures, the only way of finding out whether the method will be suitable is to try it and see. Adsorption will be encouraged by low salt concentration, but sufficient buffering should be present to avoid undesirable local pH changes in the highly dipolar environment of the adsorbent surface. Elution of proteins occurs at higher salt concentrations, which can be added directly or, in the case of column chromatography, applied as a gradient.

Some of the inorganic adsorbents that have been used are listed in Table 4.8. One often comes across the use of these materials not as chromatographic materials or even in the normal adsorption–elution cycle, but with the aim of adsorbing only unwanted material. A batch-wise or column-type system may adsorb a variety of undesirable proteins, other macromolecules, and low-molecular-weight compounds, but leave the desired enzyme in solution, immediately ready for the next step in purification. An example, using both zinc hydroxide and bentonite as batch-wise adsorbents, is shown in Table 4.9, describing the purification of yeast phosphoglucose isomerase EC 5.3.1.9 (111). Zinc hydroxide adsorbent is created in situ by adding zinc acetate plus alkali to maintain the pH around 8. Many proteins are adsorbed, but the isomerase remains in solution. After removing the hydroxide gel by centrifugation, bentonite is added, which adsorbs more proteins, but still the isomerase remains behind. Bentonite contains very fine gelatinous particles which require relatively high centrifugal force to sediment. Thses two steps achieved a 20-fold purification from a crude extract which had not been processed by any prior treatment.

**Table 4.9.** Purification of Phosphoglucose Isomerase from a Yeast Extract Using Batch Adsorption of Unwanted Proteins

|  | Protein (mg) | Enzyme activity (units) | Specific activity |
|---|---|---|---|
| Crude extract | 30,500 | 173,000 | 5.7 |
| After $Zn(OH)_2$ treatment | 6,630 | 115,000 | 17.3 |
| After bentonite treatment | 905 | 109,000 | 120 |

*Source:* Ref. 111.

## Hydroxyapatite and Calcium Phosphate Gels

The usefulness of hydrated calcium phosphate gels for selectively adsorbing proteins, allowing their release at higher salt concentrations, has been recognized and applied for decades. The original and still very useful adsorbent was prepared by mixing calcium chloride with tribasic sodium phosphate, washing the gelatinous precipitate with distilled water, then allowing it to "age" for several months before it acquired optimum, consistent characteristics (112). Being gelatinous, this material is not of any use directly in column because of poor flow characteristics, but is ideal for batch adsorption. It has a high capacity because of its large surface area per unit weight and, until the development of ion exchangers, was widely used in protein purification. Its use has been unjustifiably neglected recently with the development of sophisticated adsorbents and controlled chromatographic-quality calcium phosphate particles. Some attempts have been made to use this calcium phosphate gel mixed with Celite (a filter aid) in columns; however, there is little advantage to doing this rather than using crystalline hydroxyapatite designed for column use; the chief advantage of the gel is its use in rapid batch-wise adsorption.

To prepare hydroxyapatite, the calcium phosphate that is crystalline and suitable for columns, 0.5 M solutions of $CaCl_2$ and $Na_2HPO_4$ are slowly mixed together with a solution of M NaCl. This makes brushite, $CaHPO_4 \cdot 2H_2O$. The brushite is boiled with NaOH, which converts it to hydroxyapatite, $Ca_{10}(PO_4)_6(OH)_2$ (113).

The crystalline particles in the column adsorb proteins on their surface; unlike biologically based matrices, the proteins cannot penetrate the particles. Consequently, the adsorptive capacity is limited; indeed, a monolayer of a typical protein over the surface of spheres 0.1 mm in diameter amounts to no more than $0.1 \text{ mg cm}^{-3}$. Capacities with hydroxyapatite are higher than this, probably due to crevices and cracks within the particles giving a much higher effective surface area. But the limited capacity of crystalline calcium phosphate makes its use in enzyme purification mainly a late-stage procedure, often the final one after other techniques have failed to result in homogeneity.

Bernardi (109), in extensive studies of the behavior of proteins on hydroxyapatite, has concluded that the adsorption of basic proteins, positively charged at the operating pH, is rather different from that of neutral or acidic proteins. Adsorbed at low K-phosphate concentrations, all proteins could be eluted by increasing phosphate concentration, with basic proteins tending to require stronger buffers than acidic ones. Other salts such as KCl and NaCl had no effect in eluting acidic proteins; nor did $CaCl_2$, at up to 3 M concentrations. Basic proteins, on the other hand, were eluted by chloride salts, and very low concentrations of $CaCl_2$ (1–10 mM) were sufficient to displace such basic proteins as lysozyme and ribonuclease from hydroxyapatite. The conclusions drawn from these experiments were that acidic proteins were more adsorbed by $Ca^{2+}$ sites on the crystal surfaces, and anions such as $Cl^-$ were not effective in displacing them as they have little affinity for $Ca^{2+}$. On the other hand, basic proteins, binding more through $PO_4^{n-}$ sites, can be displaced by monovalent cations reasonably effectively, and very effectively by $Ca^{2+}$ ions which have a high affinity for the $PO_4^{n-}$ sites. Although these theoretical and model experiments have demonstrated some new possibilities for obtaining separations on hydroxyapatite columns, actual examples of using selective salt elutions are few. A gradient of phosphate buffer remains the normal procedure adopted, perhaps through ignorance of the alternatives.

A recent development from LKB is a gel bead coated with hydroxyapatite— HA-Ultrogel. This is clearly superior to other forms of hydroxyapatite in its properties, including a (relatively) high adsorption capacity. Details of applications are given in the LKB manual (114).

## Hydrophobic Adsorbents

During the development of affinity chromatographic support mediums, a number of "control" experiments were carried out, in particular looking at the behavior of matrices containing spacer arms but no ligand. In most cases these behaved as expected, with no adsorption noted. But in a few cases enzymes were found to bind strongly to hexamethylene arms, even without charged ends (amino or carboxylate) which might have been acting as ion exchangers (115). These observations were put to good use in developing the techniques of hydrophobic chromatography, for it was the interaction between aliphatic chains on the adsorbent and corresponding hydrophobic regions on the surface of the proteins that was causing binding (116,117). Indeed, it is this "nonspecific" contribution to affinity adsorption that is most helpful in achieving tight binding (cf. section 4.5).

Whereas relatively few proteins bind to short immobilized aliphatic chains at low salt concentration, hydrophobic chromatography can be extended to cover all proteins, since the hydrophobic interactions *increase* in strength with increasing salt concentration. In particular the salts which are most effective in salting-out precipitation (cf. section 3.3) such as ammonium sulfate are the ones that strengthen the hydrophobic interaction most. The reasons are the same; in salting out, a main cause of aggregation is the strengthening of hydrophobic interaction between proteins. Consequently at high salt concentrations most proteins can be adsorbed to hydrophobic groups attached to inert matrices (however, the matrix itself may not be inert; see below).

Typical hydrophobic adsorbents available commercially include $C_4$, $C_6$, $C_8$, and $C_{10}$ linear aliphatic chains (P.L. Biochemicals), and the same chains containing a terminal amino group (Figure 4.51a). The latter were intended principally for affinity ligand attachment, but in themselves are interesting hydrophobic adsorbents with a potential hydrogen- or electrostatic-bonding group to alter the behavior. It is also worth noting that these products are based on cyanogen–bromide-activated agarose, and so have a positively charged group at the other end of the aliphatic chain (cf. Figure 4.35). Phenyl-agaroses are

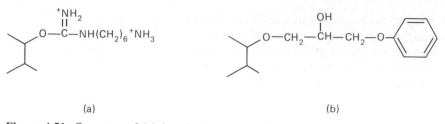

(a)                                                                        (b)

**Figure 4.51.** Structure of (a) hexylamine-agarose (P-L Biochemicals, Aghexamine) and (b) phenyl-Sepharose (Pharmacia).

also available in which a benzene ring is attached via a glycerol ether linkage to the matrix, e.g., Phenyl-Sepharose (Figure 4.51b). As with dye ligand adsorbents (cf. section 4.6), the potential for varying the nature of the substituent is infinite, but the behavior of each is similar. For trials, samples of phenyl-, octyl- or hexyl-, and aminohexyl-agaroses should cover most potential uses.

Hydrophobic adsorbents function as a result of interactions different from other adsorbents; thus in theory they have a good chance of effecting a separation of components which behave identically in, say, ion exchange chromatography. In practice the use of hydrophobic chromatography has been limited because it does not usually achieve very sharp separations. This is because the $\alpha$ value changes only slowly with altering conditions, and there is a large degree of overlap between the components (Figure 4.52). Also, since the principles are very similar to salting-out fractionation, if the latter has already been carried out there is less chance of getting a good separation on a hydrophobic column. Weighed against these pessimistic views are several positive points. First, the capacity for proteins is very high, in the same range as ion exchangers, namely from 10 to 100 mg cm$^{-3}$. Second, because adsorption is carried out at high salt concentration, it is not necessary to change the buffer of a sample before appli-

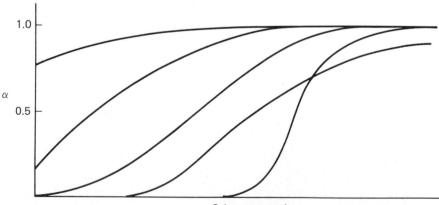

**Figure 4.52.** Variation of partition coefficient $\alpha$ with salt concentration for hypothetical collection of five proteins on a hydrophobic adsorbent.

cation; one needs only to add sufficient salt to ensure binding of the component required. And because of the stabilizing influence of salts, recoveries are often excellent. Gradients of decreasing salt concentration are sharply maintained because of density stabilization, ensuring optimum column behavior.

Proteins that are strongly adsorbed even at low salt concentration are generally those with a low water solubility: globulins, membrane-associated proteins, and others that precipitate in a low range of ammonium sulfate saturation (20–40%). These may not be readily removed from the column. Hydrophobic interactions can be weakened further by (a) lowering of temperature, (b) inclusion of organic solvent, (c) inclusion of polyols, especially ethylene glycol, (d) inclusion of nonionic detergent, (e) lowering of pH—this may be mainly a reduction of ion exchange effects due to positive charges associated with the isourea linkage to the matrix (Figure 4.35). Thus, following a decreasing salt gradient, a hydrophobic column can be developed further with increasing ethylene glycol concentration at low salt. Alternatively, after adsorption carried out at room temperature, a gradual decrease in temperature can be used as the mechanism of gradient elution.

It is fair to say that the potentials of hydrophobic adsorbents have not yet been realized. Despite its lack of chromatographic selectivity between similar components, hydrophobic adsorbents should be very valuable in cruder separations at quite early stages in the purification, even the first step. As a rapid batch technique it could replace some ammonium sulfate frationations by eliminating centrifugations and achieving a sharper cut; this is described in the section below.

## "Salting-Out" Chromatography

As mentioned above, hydrophobic chromatography and salting-out precipitation are closely related; the combination of the two is a natural extension. Adding a slurry of ammonium-sulfate-precipitated protein to a column preequilibrated with ammonium sulfate, and then developing with a decreasing salt gradient had been successfully attempted before hydrophobic adsorbents were synthesized (118). The columns consisted of cellulose or Sephadex, and later agarose (119)—neutral, "inert" materials. However, the "inertness" of these materials at high salt concentration is not strictly true; in fact, they behave very much like hydrophobic adsorbents.

Cellulose is insoluble in water largely because of its hydrophobic character. The hydrogens on glucose residues are on the crystal surfaces, with the hydroxyls in the interior forming hydrogen bonds with adjacent molecules. But specially treated cellulose, exposing more hydroxyl groups, has been shown to act very successfully as a "salting-out adsorbent" (120). Moreover, substituted ion exchange celluloses appeared to adsorb proteins even more effectively, despite the fact that at high salt concentration the ion exchange character could be expected to be lost. Proteins adsorbed to these cellulose matrices could be eluted *at the same high salt concentration* as used for their adsorption, using hydrogen-bond-weakening solutes such as glycerol, sucrose, urea, or ethanol (120). Consequently it was surmised that hydrogen bonding in these conditions was at least as important an attractive force as the hydrophobic interactions.

Agarose contains many methylene residues as well as hydroxyls, since it contains anhydrogalactan residues. The amphiphilic nature of these matrices (partly hydrophobic, partly hydrophilic) may be even more suited to protein adsorption at high salt concentrations than truly hydrophobic materials (121). Indeed, agarose does appear to be a most suitable matrix for salting-out chromatography. Proteins adsorb to the agarose when the salt concentration is somewhat less than the concentration required to precipitate them in free solution; thus it is not necessary, nor desirable, to pre-precipitate the protein and apply it as a slurry. Two techniques are possible: (1) Add ammonium sulfate to a concentration just below that needed to precipitate the enzyme of interest, centrifuge off any precipitate, and apply supernatant to column preequilibrated at the same ammonium sulfate concentration. (2) Add agarose to sample, approximately 1 cm$^3$ per 20 mg protein, and slowly dissolve in ammonium sulfate until the concentration (based on the whole volume of agarose + sample) is about that required to precipitate the protein from free solution. Then transfer agarose to top of partly poured column as above. In either case the proteins, rather than self-aggregate, adsorb to the agarose and coat its surface (internal as well as external) with protein, up to 30 mg cm$^{-3}$. After washing with starting buffer, the gradient, or stepwise lowering of salt concentration can begin. This, in principle, is the same as hydrophobic chromatography, but it is likely that no proteins will remain adsorbed at low salt concentration. The poorer hydrophobic character of agarose is compensated for at the adsorption stage by using a higher salt concentration.

An extension to this procedure involves an "affinity" technique, in which the presence of a ligand in the buffer alters the adsorption characteristics (122). Either with hydrophobic adsorbents or on agarose, the principles of affinity methods remain the same as with more specific adsorbents (cf. section 4.5) or ion exchangers (cf. section 4.4). First one establishes conditions without ligand which are not quite sufficient to cause the effect, e.g., in this case, elution. Then addition of the ligand just tips the balance and causes the effect. It is essential that the ligand makes a detectable difference to the property concerned, in this case the hydrophobic interaction. It must be able to bind to the enzyme in the conditions on the column, which involves a high salt concentration. The example of "affinity precipitation" presented is for some amino tRNA transferases (122); binding of the large tRNA molecule causes a sufficient decrease in adsorption properties to allow elution.

## Protamine–Nucleic Acid Complexes as Adsorbents

The first stage in cleaning up extracts of microorganisms, and other fast-proliferating tissues which contain large quantities of nucleic acids, is usually to precipitate these nucleic acids with protamine sulfate (cf. section 2.3). The resultant precipitate usually contains little protein other than nucleoprotein, and the process simply removes undesirable material from the extract. However, it has occasionally been observed that a particular enzyme activity is decreased substantially after a protamine precipitation, and the enzyme lost subsequently has been recovered by extracting the precipitate with some suitable buffer. When this phenomenon arises, it can be used as an important means of purification. Thus *E. coli* α-ketoglutarate dehydrogenase adsorbs to the protamine–nucleic acid precipitate, and is extracted by 0.1 M phosphate pH 7 (123). Recently this phenomenon was used for purifying yeast phosphofructokinase which was totally adsorbed on a protamine precipitate (17). Since these examples are of enzymes involved with nucleotides, the adsorption may be through exposed insolubilized nucleotide ends on the nucleic acid.

## Mixed-Function Adsorbents

In the decade since dipolar ion exchangers were first described (69) little use has been made of them despite the fact that their described properties seem useful. Sulfanilic acid and arginine were coupled to epichlorhydrin-activated agarose (Sepharose 6B) to give dipolar substituents (Figure 4.53). The secondary amine $pK_a$ values were found to be 7.25 (sulfanilic acid–Sepharose) and 8.10 (arginine–Sepharose), so the structures shown in Figure 4.53 presumably apply to pH values below these $pK_a$ values. Selective adsorption of plasma proteins was demonstrated, and protein adsorption capacities were as high as those of conventional ion exchangers. The sulfanilic acid adsorbent may have shown some hydrophobic characteristic also; it could be regarded as a highly simplified structural analogue of a dye ligand (cf. section 4.6).

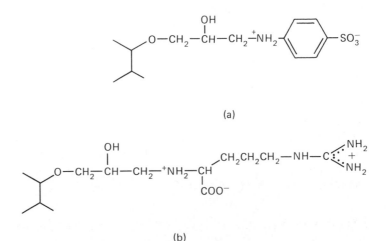

(a)

(b)

**Figure 4.53.** Dipolar "ion exchange" adsorbents based on agarose gel: (a) sulfanilic acid–agarose and (b) arginine–agarose. However, it is unlikely that the sulfanilate nitrogen in (a) would be protonated above pH 4. [From Porath and Farnstedt (69).]

A similar development of mixed-function adsorbents has been undertaken by Yon (64–66). In this case hydrophobic and hydrophilic character was introduced to the "affinity" ligand, and good selectivity was observed. However, very low adsorption capacities have restricted the use of his materials.

## Phenyl Boronate Adsorbents

A recent development applicable to enzyme purification is the phenyl boronate–agarose adsorbent. This operates by a reversible covalent linkage between the immobilized boronate and cis-hydroxyl groups on soluble compounds (Figure 4.54). The adsorbent can thus act as a general affinity adsorbent selecting for adjacent hydroxyl groups. However, proteins only have such hydroxyls if they contain prosthetic groups, notably carbohydrates. Thus the phenyl boronate adsorbents provide an alternative method for binding glycoproteins compared with the more traditional lectin–agaroses.

A more interesting application has been called "piggyback chromatography" (124). In this case the column is preloaded with ligand containing the hydroxyls, e.g., a sugar, or a ribonucleotide such as ATP or NAD. A high density of ligand is loosely (reversibly) coupled to the column ($m_t$ approaches $10^{-1}$ M for Eq. 4.5). Then the sample is put on, and the enzyme, binding to the ligand, is retarded. As the coupling is reversible, a proportion of the ligand moves down at any given moment, and the enzyme "rides piggyback" on the slowly mobile ligand. The ligand movement may be so slow that excess ligand (or a different one) may be needed to elute the enzyme.

Because of the high value of $m_t$, a high $\alpha$ is achievable even with a low affinity system; e.g., $m_t = 3 \times 10^{-2}$ M, $K_P = 5 \times 10^{-3}$ M, and $p_t = 10^{-4}$ M gives $\alpha = 0.86$, a respectable retardation, and highly specific for proteins binding to the ligand in question. Unfortunately, there can be some nonspecific

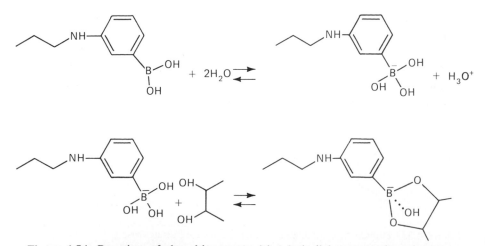

**Figure 4.54.** Reactions of phenyl boronate with 1,2-cis-diol compounds such as may be found on the carbohydrate moiety of glycoproteins and on the ribose ring of nucleotides. [From Amicon (124).]

attraction to the phenyl ring which can confound the separations. Further details of the methods are explained in the booklet from Amicon Corporation (124).

## 4.9  Batch Adsorption

Some of the simplest steps in enzyme purification make use of batch adsorption, in which a flocculant material is added to a solution, often the initial extract, and proteins are adsorbed to it. Any adsorbent can be used; it should be capable of rapid recovery and washing by flocculation or filtration; on smaller scales centrifugation may be appropriate. The chief requirement for a successful batch method is that the partition coefficient $\alpha$ of proteins being adsorbed should be very close to 1 and that a significant number of other proteins should have lower $\alpha$ values. The reason for this is indicated in Figure 4.55 and Table 4.10, where it is seen that the *concentration* of free protein is the same throughout the volume, so the total *amount* not adsorbed can be substantial if $\alpha$ is even fractionally below 1.

Figure 4.55 allows the following calculations. For a two-phase system: free liquid and settled adsorbent volumes are $V_s$ and $V_a$, respectively. The concentration of free protein, $p$, in $V_a$ is same as in $V_s$. The amount of protein adsorbed is $qV_a$, where $q$ is the concentration of adsorbed protein. The amount of protein not adsorbed is $p(V_a + V_s)$. The partition coefficient $\alpha$ is defined as $\alpha = q/(p + q)$. This allows the concentration of free protein to be written as $p = q(1 - \alpha)/\alpha$. Therefore, the fraction protein adsorbed, $f$, is

$$f = \frac{qV_a}{p(V_a + V_s) + qV_a} = \frac{V_a\alpha}{V_a + (1 - \alpha)V_s} \tag{4.11}$$

Table 4.10 gives an idea of how close $\alpha$ must be to unity to obtain a good adsorption yield of 80–90%. With most adsorbents, $K_P$ values smaller than $10^{-6}$ M are needed to give $\alpha$ values of 0.98 or greater. If $\alpha$ is over 0.99, very small amounts of adsorbents can be used; a value of $V_a/V_s$ of 0.02 would not be unreasonable if the enzyme concerned is only present in low amounts. However, it is uncommon for general adsorbents to be so specific; large amounts of

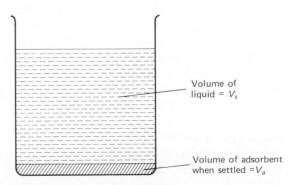

**Figure 4.55.** Definition of volumes in batch-wise adsorption procedure.

Volume of liquid = $V_s$

Volume of adsorbent when settled = $V_a$

**Table 4.10.** Fraction of Protein ($f$) Adsorbed
in Different Batch Conditions, from Eq. (4.11)

| $V_a/V_s$ | $\alpha$ | $f$ |
|-----------|----------|-----|
| 0.05 | 0.9 | 0.30 |
| 0.1 | 0.9 | 0.45 |
| 0.2 | 0.9 | 0.60 |
| 0.5 | 0.9 | 0.75 |
| 0.1 | 0.95 | 0.63 |
| 0.2 | 0.95 | 0.76 |
| 0.5 | 0.95 | 0.86 |
| 0.05 | 0.98 | 0.70 |
| 0.1 | 0.98 | 0.82 |
| 0.1 | 0.98 | 0.89 |
| 0.1 | 0.99 | 0.99 |
| 0.02 | 0.998 | 0.91 |
| 0.05 | 0.998 | 0.96 |

other proteins are likely to bind, reducing the effective capacity of the adsorbent. Affinity adsorbents are much more likely to be selective, but, as mentioned in section 4.6, $K_P$ values smaller than $10^{-6}$ M are unlikely to be achieved without substantial nonspecific interactions. But if an affinity adsorbent is based on a ligand that binds very tightly to the enzyme, and there is little nonspecific interaction (through using a hydrophilic spacer arm, or by direct attachment of the ligand to the matrix), then batch treatment can be highly successful. Immunoadsorbents should be particularly suitable for batch applications, since they select out the antigen and bind it tightly. Dye ligand adsorbents fulfill these conditions also, but are relatively nonspecific; consequently quite large amounts may be needed before all the enzyme wanted is adsorbed. As an example, glucose 6-phosphate dehydrogenase (from yeasts and bacteria) can be 90% adsorbed onto Procion Red HE-3B-Sepharose from a crude extract at $V_a/V_s = 0.1$. From Table 4.10 this implies an $\alpha$ value of 0.99 or higher.

Sometimes one must judge whether a batch treatment is better than a column. Batch methods offer the advantages of speed and the possibility of treating large volumes of material, but unless $\alpha$ is well above 0.98, some losses are inevitable. Columns can result in 100% adsorption even at $\alpha$ values of 0.95, but are more trouble, slower, and may clog if dealing with crude extracts. Consider the situation of having 1 liter of material and adding 100 cm$^3$ of adsorbent, resulting in 63% of the enzyme being adsorbed ($\alpha = 0.95$, Table 4.10). If this $\alpha$ value is applicable to a column, and the capacity ($m_t$) is far greater than the amount of enzyme present, then running the whole sample on to a 100-cm$^3$ column will result in the leading edge of nonadsorbed protein being located halfway down [movement rate = flow rate $\times$ $(1 - \alpha)$, volume passed in = 1 liter, so movement = 50 cm$^3$]. Washing with a further 500 ml of buffer will still not cause elution of any enzyme, and a stepwise or gradient elution procedure now should result in an excellent chromatographic separation. But if the procedure is to be scaled up to, say 20 liters, then columns may not be feasible. The question is whether 63% adsorption by a batch treatment is a sufficiently good recovery; some more may be lost on washing. The answer is probably no—$\alpha$ must be increased somehow or the amount of adsorbent increased. Using twice as much adsorbent increases the recovery to 76% (Table 4.10), which is more acceptable. Perhaps diluting the solution with water, thereby halving the ionic strength, will increase $\alpha$. But using the same amount (100 cm$^3$) of adsorbent, even if $\alpha$ increases from 0.95 to 0.98, the recovery only goes to 70% at the adsorption stage because of the large volume ($V_a/V_s$ becomes 0.05).

Even if the enzyme does not adsorb at all, batch processes can be extremely useful for removing other proteins. An example was given in the previous section, in which first zinc hydroxide and then bentonite was used to adsorb much of the protein material from a yeast extract, but leaving all the phosphoglucose isomerase in solution (111) (Table 4.9). There are many other examples, sometimes going under the category of "clarification," but incidentally involving adsorption of unwanted material. Even if the degree of purification is not great (e.g., only 10–20% increase in specific activity), the protein removed may otherwise have been an important contaminant at a later stage.

## General Approaches

The simplest type of batch adsorption involves adding the adsorbent (preequilibrated with the appropriate buffer), stirring for a few minutes, allowing to settle, and then filtering under suction. The pad on the filter funnel is then washed with buffer until little nonadsorbed protein remains. It is sucked to damp-dryness, then the elution buffer mixed in (temporarily removing the suction) and allowed to stand with the adsorbent for a few minutes. It may be better to transfer the adsorbent with the elution buffer to a beaker and stir for a few minutes, before returning the slurry to the filter funnel. Meanwhile the collecting flask is washed out thoroughly and drained, before replacing it and sucking the elution buffer through. The process is then repeated, and finally, without mixing the pad, a further volume of buffer passed in to ensure that all the desorbed enzyme is washed out. Each batch of elution buffer should be kept to only 1–2 times the volume of the adsorbent pad. The process is described diagrammatically in Figure 4.56.

More complex protocols may lead to better purification. For instance, there may be proteins present which adsorb more strongly than the enzyme required, so none of the enzyme binds to the first bit of adsorbent added. In this case the first batches of adsorbent should be removed before adding the batch that takes out the protein (Figure 4.57). A small-scale trial to find the optimum batches of adsorbent can be carried out first.

Elution can also be stepwise, removing some unwanted material in an initial elution before the buffer to elute the wanted enzyme is added. Again a small-scale trial can find the optimum (Figure 4.58).

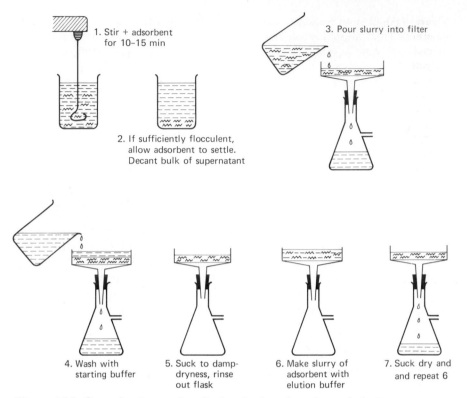

1. Stir + adsorbent
   for 10–15 min

2. If sufficiently flocculent,
   allow adsorbent to settle.
   Decant bulk of supernatant

3. Pour slurry into filter

4. Wash with
   starting buffer

5. Suck to damp-
   dryness, rinse
   out flask

6. Make slurry of
   adsorbent with
   elution buffer

7. Suck dry and
   and repeat 6

**Figure 4.56.** Operational procedure for batch-wise adsorption and elution.

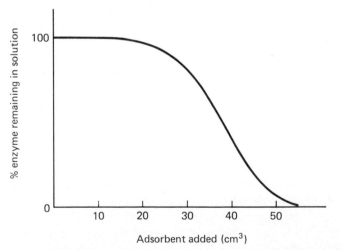

**Figure 4.57.** Measurement of enzyme activity remaining in solution after successive batches of adsorbent indicates optimum protocol. In this example, 20 cm³ of adsorbent should be added first and then removed. Then a further 30 cm³ is added to adsorb the enzyme.

**Figure 4.58.** Stepwise elution to determine optimum procedure for extracting enzyme. In this example enzyme is eluted mostly above 20 mM salt. Probably the optimum protocol would be (i) wash with 20 mM salt to remove unwanted proteins, (ii) wash two times with 50–60 mM salt to obtain most of enzyme without too much nonspecific protein.

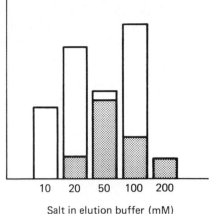

Salt in elution buffer (mM)

Typical batch adsorbents are:

    Calcium phosphate gel
    Ion exchangers (especially phosphocellulose)
    Affinity adsorbents
    Dye ligand adsorbents
    Hydrophobic adsorbents
    Immunoadsorbents

In the category of hydrophobic/salting-out adsorption are procedures for separating protein from other undesirable materials, such as microorganism capsular gum and other carbohydrates—this has been discussed in Chapter 2.

Because of the speed and potential large-scale use of batch methods, they should be seriously considered in any preparation where an adsorbent is used. With so many different types of adsorbent being produced, e.g., a multitude of dye ligand possibilities, it becomes more likely that one exists with the appropriate characteristics for any particular enzyme. If it can be applied to a crude extract, there is a great saving in time, and requirements for larger-scale equipment may disappear.

## General References

Amicon Corporation publications (1980) *Dye-ligand chromatography: Applications, method, theory of Mātrex gel media,* Lexington, Mass.

T. C. J. Gribnau, J. Visser, and R. J. F. Nivard (eds.) (1982), *Affinity chromatography and related techniques.* Elsevier, Amsterdam.

W. B. Jackoby (ed.) (1974), Affinity techniques: Enzyme purification, part B. In *Methods in Enzymology,* Vol. 34, Academic Press, New York.

C. R. Lowe (1979), An introduction to affinity chromatography. In *Laboratory techniques in biochemistry and molecular biology* (T. S. Work and and R. H. Burdon, eds.), Vol. 7, Part II, Elsevier/North-Holland, Amsterdam.

C. R. Lowe and P. D. G. Dean (1974), *Affinity chromatography*. Wiley, London.

E. A. Peterson (1970), *Cellulosic ion exchangers* (T. S. Work and R. H. Burdon, eds.), Vol. 2, Part II, Elsevier/North-Holland, Amsterdam.

Pharmacia Fine Chemicals AB publications (1979), *Affinity chromatography: Principles and methods*. Uppsala.

Pharmacia Fine Chemicals AB publications (1980), *Ion exchange chromatography: Principles and methods*. Uppsala.

J. Turková (1978) *Affinity chromatography*. Elsevier, Amsterdam.

# Chapter 5
# Separation in Solution

The methods described in Chapters 3 and 4 involve phase changes. The proteins pass from liquid phase (i.e., dissolved) to solid (precipitated or adsorbed) and back again. Not all proteins withstand the stresses occurring in these methods; gentler methods in which the proteins remain in solution at all times are available. One of these, gel filtration, is one of the principal techniques used in purifying enzymes and other proteins, and it will be considered in detail. Other methods, grouped together as electrophoretic techniques, are less widely used, for reasons which will be outlined in sections 5.2 and 5.3. A third, little-used method of separation in solution involves phase partitioning where proteins may move from one liquid phase into another (section 5.4). As a general technique, separation in solution is an important procedure both in research and industrial enzyme and protein purification.

## 5.1   Gel Filtration

Several other names for this method have been put forward, including *gel permeation* and *molecular sieving.* However, the use of the term *gel filtration* is now so widespread and generally recognized that it will be used here. This name is unfortunate, since the procedure does not depend on the material used being a true gel, and it is not really filtration. Originally introduced by Porath and Flodin (125), the dextran-based materials of Pharmacia chemicals, Sephadex gels, quickly acquired a reputation as a gel filtration medium for rapid separation of macromolecules based on size. These cross-linked dextran beads soon achieved widespread use both for protein separation and for protein molecular weight determination (126,127). Since then the method has been developed and extended by improving the gels to give faster processing and to cover a larger range of molecular sizes. The basic principles are simple. The gel consists of an open, cross-linked three-dimensional molecular network, cast in bead form for easy column packing. The pores within the beads are of such sizes that some are not accessible by large molecules, but smaller molecules can penetrate all pores.

The inaccessibility may be due to the fact that a particular pore is too narrow for the molecules to pass down, or that even if large enough, there is no channel from the surface by which it can be reached—this is illustrated in Figure 5.1 in two dimensions. The two-dimensional representation is misleading, since a cross-link apparently preventing passage may be bypassed by motion out of the plane of the paper. It is now accepted that under optimum flow conditions all accessible pores can be filled as the proteins pass by. Therefore, equilibrium rather than kinetic effects are involved. Any particular grade of material has a certain proportion of pores accessible to each size macromolecule, ranging from 0 to 100%. It is not possible to make a material with just one size of pore which would give a very sharp cutoff. The best gel filtration beads have pore sizes which result in 90% exclusion of macromolecules that are 5–6 times the size (volume) of those macromolecules that are excluded from only 10% of the bead volume. This means that 80% of the pores are (apparently) within a twofold range of linear dimensions for spherical particles.

Diagrams to demonstrate the theory of gel filtration chromatography are usually clear enough, but often give a false impression of the dimensions of things. Thus Figure 5.2 illustrates the principles of the technique, but the true dimensions are approximated better by Figure 5.3. Beads are in the range $10^3$

**Figure 5.1** Two-dimensional representation of pores of gel filtration material and accessibility of various areas by molecules of different sizes.

**Figure 5.2.** Simple diagram illustrating principles of gel filtration to separate different-sized molecules in a column. Large molecules are excluded from most of the available column volume, and so move rapidly through ahead of the "solvent front."

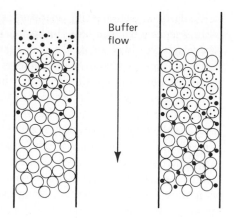

to $10^4$ times the linear dimensions of the molecules, and their surfaces on a molecular scale are of course "fuzzy."

In a column of gel filtration medium, the behavior of a particular size molecule can be expressed in a number of ways. It can be related to the total column volume $V_t$ and the void (outside the bead) volume $V_o$ by expressions such as $K_{av} = (V_e - V_o)/(V_t - V_o)$, where $V_e$ is the elution volume of the molecule being considered and $K_{av}$ is a coefficient which defines the proportion of pores that can be occupied by that molecule. Alternatively, the elution volume can be related to the elution volumes of a number of other molecules of known size, from which the molecular size of the unknown can be estimated by simple extrapolation. The elution volume relates, theoretically, to the Stokes radius, not to the size or the molecular weight. The Stokes radius describes a sphere with a hydrodynamic behavior equivalent to that of a particular irregularly shaped particle. Consequently, provided that the shapes of both the unknown and the calibrating proteins are similar, the molecular weights can be used in calibrations.

For protein purification, the exact position of elution is not critical, provided that the unwanted proteins are well separated. Usually it is desirable to arrange for the protein required to elute between one-third and two-thirds of the way down the separation range, since here the resolution is greatest; however, for particular problems other choices could be made (see below). Sephadex (cross-linked dextran) has been used for over 20 years, and will continue to find uses,

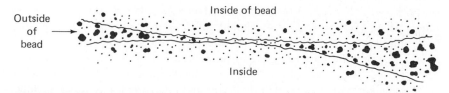

**Figure 5.3.** Scale diagram to illustrate relative sizes of molecules and beads (shaded) used in gel filtration.

particularly in commercial operation where fine judgments between speed, costs, and convenience have to be made. Cross-linked polyacrylamide beads (Biogels) have also been used widely, and cover a similar molecular size range. For very large molecules, from proteins up to virus-sized aggregates, agarose gels have been more suitable. Attempts to introduce controlled pore glass media into protein chemistry have not been so successful because of adsorption effects; these can be overcome, but the advantages of the rigidity of glass beads have largely been surpassed by recent developments. The major problem with Sephadex and polyacrylamide beads has been their softness; very gentle pressures, including osmotic pressures during chromatography, cause distortion, irregular packing, and eventually poor flow characteristics. This has largely been overcome by extra cross-linking of the dextran with acrylamide, producing more rigid beads which can be operated under higher pressure and so faster flow rates (Sephacryls, Pharmacia). The Ultrogels (LKB) are agarose gels with acrylamide polymer chains interspersed to confer greater rigidity and lower porosity than agarose itself. They are suited to most protein separations, and cover a similar fractionation range to Sephacryls S-200 and S-300. For very large molecules agarose gels are available from several suppliers. Pharmacia has introduced more rigid cross-linked agaroses, the CL-Sepharoses, but even these are likely to be superseded by their new products, Sephacryls S-400, S-500, and S-1000. A list of commercially available materials is given in Table 5.1.

It should be noted at this point that not all gel filtration materials behave completely inertly. Adsorption of particular proteins under certain buffer conditions has been noted, and indeed can be made use of as a separation procedure. Partial adsorption delays the elution so that a protein emerges as though it has a smaller molecular size than is actually the case. Adsorption can be of an ion exchange character, in which case it can be avoided by using a high-ionic-strength buffer, or it may be hydrophobic, in which case high ionic strength should be avoided. Sephacryls, which have an aliphatic chain in the cross-linker, are particularly likely to show adsorptive effects at higher salt concentrations and at low pH values. The manufacturers recommend using buffers with a pH close to neutrality.

With this great range of possible media, the beginner needs some advice as to what to purchase. Unless it is known that the enzyme of interest is of unusually low ($<15,000$) or high ($>10^6$) molecular weight, then two or three materials should suffice. Because the cross-linked dextran or agarose gels are superior in *nearly* every aspect, either the two Sephacryls S-200 and S-300, or Ultrogels AcA22, 34, and/or 44 should suffice. For small proteins, Sephadex G-75 (fine grade preferable), Ultrogel AcA54, or Biogel P-60 are equally suitable. Large proteins require agarose gels such as Sepharose 4B or CL-4B, Biogel A-5.0 m, Ultrogels A4 or A6, or the recently introduced S-400/S-500/S-1000 Sephacryl range. Although the latter also cover the range which includes a great many "typical" proteins of molecular weights in the range 50,000–500,000, the resolving power is greatly inferior to other products intended for

**Table 5.1.** Gel Filtration Media

| Company | Gel code | Gel type | Useful working range for globular molecules (MW) |
|---|---|---|---|
| Bio-Rad | P-2* | Polyacrylamide | 100–1800 |
| Biogels | P-4* | Polyacrylamide | 800–4000 |
| | P-6* | Polyacrylamide | 1000–6000 |
| | P-10* | Polyacrylamide | 1500–20000 |
| | P-30* | Polyacrylamide | 2500–40000 |
| | P-60 | Polyacrylamide | 3000–60000 |
| | P-100 | Polyacrylamide | 5000–100000 |
| | P-150 | Polyacrylamide | 15000–150000 |
| | P-200 | Polyacrylamide | 30000–200000 |
| | P-300 | Polyacrylamide | 60000–400000 |
| | A-0.5m | Agarose | $1000–0.5 \times 10^6$ |
| | A-1.5m | Agarose | $2000–1.5 \times 10^6$ |
| | A-5.0m | Agarose | $4000–5 \times 10^6$ |
| | A-15m* | Agarose | $60000–15 \times 10^6$ |
| | A-50m* | Agarose | $200000–50 \times 10^6$ |
| | A-150m* | Agarose | $1 \times 10^6–150 \times 10^6$ |
| LKB Ultrogels | AcA22 | Agarose/polyacrylamide | $60000–1 \times 10^6$ |
| | AcA34 | Agarose/polyacrylamide | 20000–400000 |
| | AcA44 | Agarose/polyacrylamide | 12000–1300000 |
| | AcA54 | Agarose/polyacrylamide | 6000–70000 |
| LKB Agaroses | A2* | Agarose | $120000–20 \times 10^6$ |
| | A4* | Agarose | $55000–9 \times 10^6$ |
| | A6* | Agarose | $25000–2.4 \times 10^6$ |
| Pharmacia Sephadex | G-10* | Dextran | 50–700 |
| | G-15* | Dextran | 50–1500 |
| | G-25* | Dextran | 1000–5000 |
| | G-50 | Dextran | 1500–30000 |
| | G-75 | Dextran | 3000–70000 |
| | G-100 | Dextran | 4000–150000 |
| | G-150 | Dextran | 5000–300000 |
| | G-200 | Dextran | 5000–600000 |
| Pharmacia Sepharoses | 6B | Agarose | $10000–4 \times 10^6$ |
| | 4B | Agarose | $60000–20 \times 10^6$ |
| | 2B* | Agarose | $70000–40 \times 10^6$ |
| | CL-6B | Cross-linked agarose | $10000–4 \times 10^6$ |
| | CL-4B | Cross-linked agarose | $60000–20 \times 10^6$ |
| | CL-2B* | Cross-linked agarose | $70000–40 \times 10^6$ |
| Pharmacia Sephacryls | S-200 | Dextran/bisacrylamide | 5000–300000 |
| | S-300 | Dextran/bisacrylamide | 10000–800000 |
| | S-400 | Dextran/bisacrylamide | $20000–2 \times 10^6$ |
| | S-500 | Dextran/bisacrylamide | multi-enzyme complexes |
| | S-1000* | Dextran/bisacrylamide | of $MW > 10^6$ |

*Not ideally suited for protein fractionation.

this range alone, due to a large relative spread of apparent pore sizes. This is a result of the inevitable flexibility of very large pores, compared with rigidity of the more cross-linked small pore gels.

Whatever grade of material is used, certain values are constant. With spherical beads it is inevitable that $V_o$ equals between 30 and 35% of $V_t$, depending on how tightly they pack in the column. Thus the useful range for resolving proteins lies within about 80% of the remaining volume ($V_t - V_o$), i.e., about 55% of $V_t$. The total volume accessible to liquid is slightly less than the total column volume, because the solid material of the gel matrix plus tightly bound water occupies a finite volume. For a column of 100 cm$^3$ total volume, one can expect resolution of proteins within the range of elution volumes 35–90 ml, with the sharpest resolution between peaks of different molecular weights eluting at around 60 ml. The actual resolution depends not only on the above features, but also on the extent of diffusion and non-ideal behavior in the column. Proteins have relatively low diffusion coefficients, but in practice much more diffusion is seen than expected from these values. This is because of non-ideal conditions even in the best-packed column. In ideal conditions a protein front would move down with the only spreading due to molecular diffusion. In practice there are two complications. First, gravitational instabilities cause the more dense solution behind the protein front to break ahead of the theoretical position (Figure 5.4). Second, turbulent flow, as liquid squeezes through the narrow spaces between the beads into the wider voids beyond, leads to additional spreading. Figure 5.5 illustrates the phenomenon of turbulent flow between the bead particles. The gravitational instability at the *front* edge of the protein band could be solved by running a column upward, but then the trailing edge, normally gravitationally stable, would break up. Only by running in zero gravity could this be solved. The turbulent flow situation cannot be countered, except by reduction in flow rate, which will alleviate but not eliminate the problem. And if the flow rate is cut back a lot, not only does it make the run inconveniently slow, but molecular diffusion effects may become larger than the turbulence.

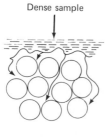

Dense sample

(a)

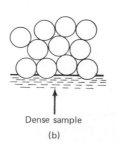

Dense sample

(b)

**Figure 5.4.** Gravitational effects at buffer-sample boundaries during application of sample to a gel filtration column. In (a), the dense sample runs ahead between the beads, causing a diffuse leading boundary. In (b), the dense sample applied in an upward flow maintains a sharp leading boundary. [From Scopes (29).]

**Figure 5.5.** Turbulent flow in gel filtration causing greater spreading of protein bands than diffusion alone.

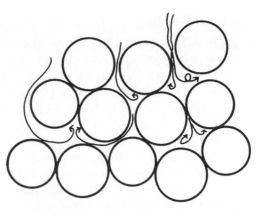

Both of these non-ideal behaviors are decreased by using smaller bead particles, resulting in the disturbances operating over shorter distances. Thus a finer grade of bead should give a better resolution. Moreover, smaller beads mean that equilibrium between outside and inside is reached more quickly because the diffusion distances are smaller, so a fairly fast flow rate can be used without losing equilibrium conditions. Unfortunately, the smaller the beads, the higher the pressure needed to force the liquid down the column, and with nonrigid beads this results in distortion and eventually blockage of flow.

The most recent advances have approached the question of resolution from two directions. By improving the rigidity of the beads it has been possible to use smaller sizes (e.g., Sephacryl S-200, 40–105 $\mu$m, compared with Sephadex G-150, 120–400 $\mu$m wet), yet much faster flow rates, up to the limit where equilibrium conditions do not have time to be attained as the proteins pass down the column. Still finer particles would need higher pressures which may not be attainable in routine work. On the other hand, the development of high-pressure liquid chromatography (HPLC) has allowed use of very fine beads (e.g., 5–10 $\mu$m), usually of a silica base, giving high resolution in a very short time. But this is mainly an analytical technique; for amounts larger than a few milligrams of sample, the size of column required necessitates reduction in pressure. Presumably the two approaches will meet in the center with moderate-pressure liquid chromatography using very fine beads, but less frenetic flow rates.

The new "Blue Column" introduced by LKB was intended for use at high pressures, but it has been found (see LKB literature) that low pressure coupled with slow flow rates gives much better resolution. The principle advantage that this material seems to have is a very wide molecular separation range; the resolution within a given range of molecular weight is little superior to that achieved with Ultrogel or Sephacryl beads with a smaller separation range.

Meanwhile, the present HPLC protein columns can deal with a few milligrams and represent an excellent procedure for the final stage of purification of a protein which makes up a small percentage of the original material.

## General Practice

For the beginner, the range of possibilities in gel filtration is much smaller than with ion exchange chromatography. The buffer used, within reasonable limits, *should* make no difference in the resolution achieved, and the only real variable is the grade of material, that is, the effective fractionation range that the material is designed for. Assuming that one has two or three of the more easily handled cross-linked dextran or agarose gels (Sephacryl, Ultrogel), it is just a matter of choosing which to use and what size column is needed. The sample is applied in a small volume, because the volume containing each component protein will keep increasing due to the diffusion and non-ideal flow effects discussed above. Since the effective fractionation volume is approximately half the volume of the column, sharpness of resolution relies on application of the sample in a small volume. This should not exceed 3% of the total column volume; smaller volumes than this will give slightly better results, down to about 1% of the column volume. Below this value diffusion and non-ideal flow will result in much the same spreading, however small the sample may have been (Figure 5.6).

On elution, 90% of each component is typically contained in a volume representing about 8–10% of the column volume, the spread depending somewhat on the elution position. The first fractions to be eluted, the largest-size molecules, have spent less time in the column and have been subjected to less turbulence. Consequently the spreading is less than fractions eluted later. Moreover, the smaller molecules have higher diffusion coefficients, so that they

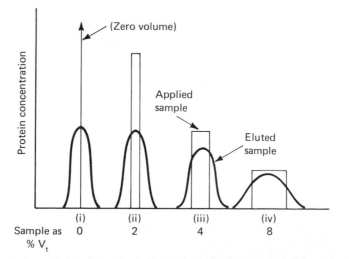

**Figure 5.6.** Effect of sample volume on elution pattern from a gel filtration column. Despite the theoretically zero volume in (i), diffusion spreads the peak to a volume similar to (ii). In (iii) the sample volume is larger than ideal, resulting in a larger total elution volume. In (iv) the sample volume is much too big except for separations of molecules differing in molecular size by a factor of at least 4. [From Scopes (29).]

spread out still further. An idealized representation of these situations is shown in Figure 5.7. If the desired enzyme is in the lower size range, it will spread more and will have more chance of being contaminated by similar-size proteins. If it elutes early, the band will be sharper, but if a lot of larger-size unwanted protein is present, this has more chance of overlapping with the enzyme fraction (Figure 5.8). Thus the choice of fractionation range must be made with consideration of the sizes of the main contaminants as well as the desired enzyme.

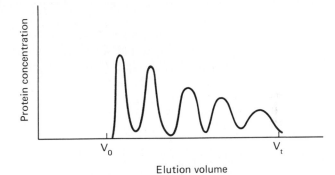

**Figure 5.7.** Ideal separation of five proteins, molecular weights differing each by a factor of 2, on a gel filtration column. The same amount of each protein was applied to the column; the peak heights decrease because of greater diffusion of the smaller molecules. [From Scopes (29).]

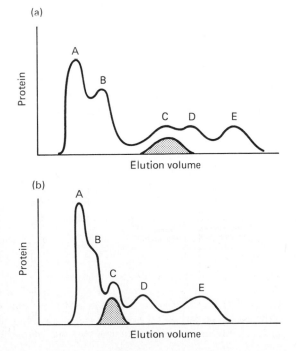

**Figure 5.8.** Separation of one enzyme (shaded) from other proteins on gel filtration column, using different-pore-sized materials: (a) large pores—the enzyme is eluted late, and is well separated from proteins A and B, but not from D; (b) small pores—the enzyme is eluted close to the void volume, and overlaps with protein B; it is well resolved from protein D.

The column dimensions are important, shape as well as total volume. In ideal conditions the shape would not matter much; indeed, a squat column would have the advantage of a faster flow rate and less turbulence, leading to less spreading (cf. Figure 1.7). On the other hand, the resolution might be less because the number of beads passed, and so the number of gel filtration events, as the proteins move in and out of the beads, for each molecule would be less. But the main disadvantage of a squat column is the difficulty of designing and operating it so that completely homogeneous sample application and buffer flow are achieved. Very minor deviations from the perfect disc containing each component would lead to poor resolution (Figure 5.9).

A long, thin column is unlikely to suffer so much from horizontal plane distortion, and such distortion would matter less anyway, since the sample would occupy a greater column depth (Figure 5.9). In practice most routine gel filtration chromatography is carried out in columns 20 to 40 times longer than their diameter. Irregular sample application, surface tension drag on the sides of the column, and other causes of irregular flow then become less important than if a short, squat column (of the same total volume) were used.

The compactness of a protein band can depend on the mode of sample application, discussed below. Equally important is attention to the fate of the protein band as it emerges from the column. Longitudinal mixing in tubing that is too wide is a common mistake, spoiling many excellent separations. Similarly, any flow cell for monitoring the protein must have small dimensions to minimize such mixing, and must be placed as close as possible to the outlet from the column (Figure 5.10).

The column should be poured in a single step if possible, using a suspension of no more than 2 times the settled volume, so that there can be little separation of large and small beads during the settling (cf. Figure 1.11). This packing should be as fast as possible without compression; a good way of creating a higher flow rate is to attach a long piece of tubing (not too narrow) to the bottom of the column so as to increase the total liquid height; the pressure drop across the column itself is then greater.

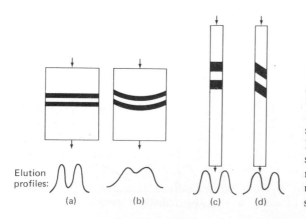

**Figure 5.9.** Comparison of gel filtration in a squat column and a long column. Ideal behavior (a) and (c) should give identical results. Minor deviations in flow in squat column (b) gives poor resolution, but in long column (d) they do not matter so much.

Elution profiles:

(a)    (b)    (c)    (d)

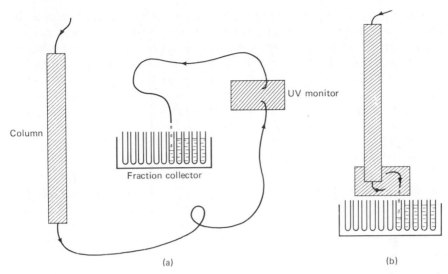

**Figure 5.10.** Incorrect (a) and correct (b) ways of arranging connections from a gel filtration (or any other) column to protein monitor and fraction collector. The length of tubing should be kept to a minimum, and tubing should be as narrow as possible without restricting the flow.

Sample application is a most important consideration. The more evenly it is applied, the better the final result. Well-designed sample applicators creating an even flow across the whole surface of the gel medium can be used successfully (cf. Figure 1.8). Do not allow any substantial dead space between the top surface and the applicator, as this will create mixing (cf. Figure 1.10). Avoid applying a sample that is less dense (e.g., containing acetone) than the buffer, since the follow-up buffer will tend to run below the last bit of sample, and this last bit then trickles in slowly long after the remainder is all into the gel. Test an applicator by using a colored sample; if uneven sample bands are observed, manual application is advisable. Remove the applicator, allow the buffer to drain to the gel surface, close a clamp on the outlet tubing, then apply the sample with a Pasteur pipette gently, allowing it to run 5–10 mm down the side of the column, and at the same time move the pipette around the column (Figure 5.11). When all the sample is on, open the clamp and let the sample run in to the surface. Then repeat the process with buffer (about the same volume as the sample)—allow it to all run in, add more buffer, then fit in the applicator for a continuous buffer supply.

Remember that the buffer that the enzyme will be eluted in is the buffer present in the column before starting; it does not matter what the sample is washed in with or eluted with. So if the desired enzyme needs to be eluted in a special, rather expensive buffer, don't waste it. Wash in 1 column volume of the special buffer prior to application of the sample. Then follow the sample through the column using a cheap convenient solution.

Gravitational instabilities during sample application (Figure 5.4) are par-

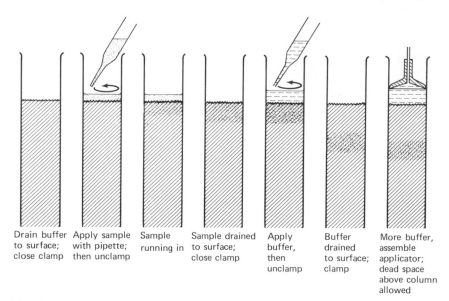

Drain buffer | Apply sample | Sample | Sample drained | Apply | Buffer | More buffer,
to surface; | with pipette; | running in | to surface; | buffer, | drained | assemble
close clamp | then unclamp | | close clamp | then | to surface; | applicator;
| | | | unclamp | clamp | dead space
| | | | | | above column
| | | | | | allowed

**Figure 5.11.** Manual application of sample to surface of gel filtration column. Care must be taken not to disturb the flat surface of the gel.

ticularly marked when a very dense solution such as a redissolved ammonium sulfate precipitate is applied to the column. The problem is greatest when using soft gels such as Sephadex G-200. With the modern rigid bead types the firmness of the surface helps to prevent excessive mixing as the dense solution layers onto the less dense buffer. Nevertheless better resolution, due to maintenance of a tight band of sample, can be obtained by applying the sample upward, at the bottom of the column, then inverting the column so that less dense buffer layers onto the surface (Figure 5.12). It is not always convenient to do this, but it would be worth trying if resolution is not as good as expected.

Finally, here is a summary of the practical aspects. Column size should be between 30 times and 100 times the volume of the sample. The starting protein concentration should ideally be in the range 10–20 mg ml$^{-1}$, 30 mg ml$^{-1}$ at the most. Thus 100 mg of protein should be in a volume of not less than 3.5 ml, preferably about 5–6 ml, and a column of volume 200–250 cm$^3$ would be ideal. The length of the column should be about 20–40 times its diameter, so in this case a column of 2.5 cm diameter and 50 cm in length would be suitable, or one 1.8 cm by 100 cm. One can determine an ideal column size as follows: diameter $= \sqrt[3]{m/10}$ cm, where $m =$ amount of protein in mg and the length $= 30 \times$ diameter.

Flow rates are restricted to the maximum attainable under a buffer head. Modern rigid materials can sustain a flow rate that is too fast for equilibrium conditions to be attained, so some cognizance of the manufacturer's recommendations is needed. Flow rates of up to 30 cm hr$^{-1}$ (30 ml hr$^{-1}$ cm$^{-2}$) are suitable for Sephacryl, but slower rates are advised for Ultrogels (also see

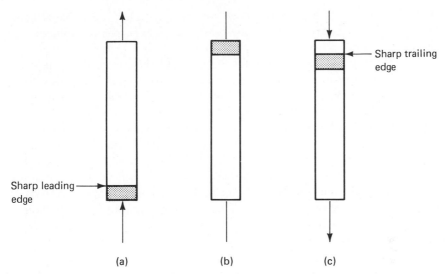

**Figure 5.12.** Scheme to avoid gravitational instabilities as illustrated in Figure 5.5: (a) the dense sample is applied to the bottom of the column with upward flow; (b) when the sample has all been applied, the column is inverted; (c) less dense buffer follows the sample—the column is run with downward flow.

Chapter 7). The time of separation can be calculated from the column dimensions. In the example above, in a column 2.5 cm in diameter and 50 cm long, if the enzyme was eluted in the middle of a Sephacryl separation range, it would emerge about 2 hr after application (at a flow rate of 30 cm hr$^{-1}$). For the $1.8 \times 100$ cm column, the time would be twice as long, though there is a chance of slightly better resolution with the longer column. As indicated earlier, finer, more rigid materials allow even faster separation under pressure, and it is likely that moderate-pressure techniques will become common, falling between routine low-pressure operation and high-pressure rapid procedures using HPLC equipment.

## 5.2 Electrophoretic Methods—Simple Electrophoresis

Separation of protein mixtures by electrophoresis began early in the twentieth century. But until relatively recently the technique was only suited for analytical purposes. Complete separation of components is not usually accomplished in the complex free-solution Tiselius apparatus which was so widely used for 30 years or more. It is still true to say that electrophoresis is principally an analytical technique. All the subtle developments of gel electrophoretic methods, isoelectric focusing, and isotachophoresis have been primarily aimed at improved analysis of protein mixtures. Nevertheless, each is capable of being adapted to a preparative scale, using tens or even hundreds of milligrams of protein mixture.

Electrophoretic steps are not often employed in enzyme purification procedures. The reasons are many, and include the complexity and expense of suitable equipment and the difficulty of operating it. A second, but not insignificant, consideration is that electrophoretic separation depends mainly on differences in charge/isoelectric points, a principle that is well exploited in ion exchange chromatography. Only in the high-resolution methods of gel electrophoresis (exploiting size differences as well) and isoelectric focusing is electrophoresis capable of achieving separation that cannot easily be obtained in one step by any other method. But gel electrophoretic methods have a variety of problems, as will be outlined below. Thus, while a preparative electrophoresis system can be a valuable asset to the enzyme purification laboratory, beginners are advised not to consider it.

## Electrophoretic Principles

A protein molecule in solution at any pH other than its isoelectric point has a net average charge. This causes it to move in an applied electric field. The force is given by $E$, the electric field ($V\,m^{-1}$) times $z$, the net number of charges on the molecule. This force is opposed by viscous forces in the medium (just as in centrifugation; cf. section 1.2), proportional to the viscosity $\eta$, particle radius $r$ (Stokes radius), and the velocity $v$; in a steady state:

$$Ez = 6\pi\eta rv \tag{5.1}$$

The specific mobility $u = v/E$ is given by

$$u = \frac{z}{6\pi\eta r} \tag{5.2}$$

It is sometimes stated that electrophoresis in free solution separates molecules according to charge alone, independent of size. But from Eq. (5.2) it can be seen that the mobility $u$ is inversely related to the Stokes radius. So a spherical molecule of 20,000 daltons, charge $-5$, should have a different mobility to one of 40,000 daltons, charge $-5$, but the same as one of 160,000 daltons, charge $-10$. But in practice the resolution in free solution would barely separate the first two of these proteins. This because of the technical problems of maintaining stable boundaries and the considerable diffusion that occurs in a typical run. Free boundary (Tiselius) electrophoresis, operated as an analytical system, rarely produces more than about eight discernable components even from the most complex mixtures because of diffusion, overlapping of components, and sometimes protein–protein interactions. Interactions between different proteins at low ionic strength may be of many types; electrostatic forces are the more important for many albumin-type, i.e., highly soluble proteins. The electrostatic interactions can be minimized by increasing the ionic strength (cf. discussions of salting in, section 3.2, and salt elution from ion exchangers, section 4.2), but an increase in ionic strength results in a higher current in electrophoresis; consequently there is more heating.

A major problem in designing a preparative electrophoresis apparatus is to allow for efficient heat removal, so that all parts of the system remain at a constant temperature. Thus a compromise must be made between high salt concentration to minimize protein-protein interactions and low salt concentrations to lessen heat output. Often all the salt is present as buffer (although a strong buffer is not really necessary, provided that electrolysis products are kept well away from the proteins). Low-conductivity buffer mixtures can be employed, in which the ionic components are relatively large and have low electrophoretic mobility e.g., $HTris^+$. As buffers these ions have extra low mobility because part of the time they are uncharged, e.g., $borate^- \rightleftharpoons H\text{-}borate$. Thus Tris-borate for the pH range 8–9 has been a popular buffer for electrophoresis. At pH 7–8, Tris-Mops (Mops = morpholinopropane sulfonate) is also a recommended low-conductivity buffer. Electrophoresis is normally carried out at neutral or slightly alkaline pH, when most proteins migrate toward the anode.

Gel electrophoresis separates molecules of otherwise similar electrophoretic mobility if they have different molecular sizes. Originally a gel of starch was employed (128), and starch gels are still very popular for analytical purposes. But starch gel has never been used successfully in a preparative way. Polyacrylamide gels (129) have become a universal feature of biochemical laboratories for protein (and now nucleic acid) analysis, and are also suitable for preparative electrophoretic techniques. Because the gel, a three-dimensional network of filaments forming pores of various sizes (130), presents a different effective viscosity depending on the size of the molecule, Eq. (5.2) for electrophoretic mobility, $u = z/6\pi\eta r$, becomes a more complex function. Within the "molecular sieving range" the mobility is linearly related to the log of the molecular weight for similarly shaped molecules (131), and in the presence of dodecyl sulfate (section 9.1) (132), $u = u_0(A - \log MW)/A$, where $A$ is the log of the molecular weight of a molecule that would not move in the gel if the extrapolation extended linearly over the whole range (Figure 5.13). Since $u_0$, the mobility of a small molecule unaffected by the gel, equals $z/6\pi\eta r$, then

$$u = \frac{z}{6\pi\eta r} \frac{A - \log MW}{A}$$

and we can regard the expression $A\eta/(A - \log MW)$ as the effective viscosity in the middle range.

Gel electrophoresis separates on the basis of both charge and size; the gel can be designed to give pore sizes suitable for the particular set of proteins being separated. A major advantage of gels is their stabilizing effect in minimizing convectional and diffusional movements of the proteins. Consequently a sharp band remains sharp as it moves through the gel, so that complete resolution of proteins of very similar mobility can be achieved. These effects will be discussed further in Chapter 9. For analytical work gels are excellent; for preparative work, although potentially excellent, they have major problems.

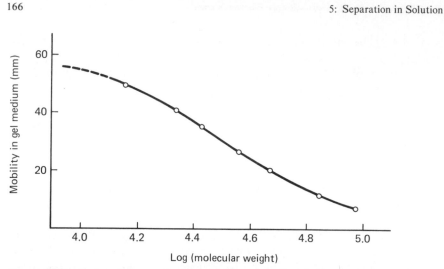

**Figure 5.13.** A plot of electrophoretic mobility versus log of molecular weight for similarly shaped protein molecules of the same charge density. Very small molecules all have the same mobility as the gel presents no extra viscous resistance.

## Methods for Preparative Electrophoresis

### Horizontal

The requirements for successful electrophoresis are a separation channel, which may be a column (vertical), a flat bed, or a rectangular block (horizontal) in which the proteins move; a connection at each end of this channel via large buffer reservoirs to the electrodes, and a cooling system to maintain the channel at a constant temperature. Because of the difficulties in maintaining stable boundaries in free solution, the buffer in the channel must be mixed with an inert powder or bead type of material to minimize gravitational and diffusional instabilities. The simplest system is the horizontal block; dry starch powder, Sephadex G-25 beads (the proteins being excluded), or powder of a variety of other polymeric compounds is mixed with the buffer to form a thick slurry which is poured into the apparatus; excess buffer is drained off. Alternatively, a very open-pore gel, e.g., 2% agar or agarose, is set in the apparatus. The ends may be supported by a wettable cloth or other material, making contact with the reservoirs of buffer which lead to the electrodes (Figure 5.14).

A vertical slot is cut in the block and refilled with a slurry of the powder suspended in the protein sample (which has been preequilibrated in the same buffer). The current is then switched on and electrophoresis allowed to proceed. If the apparatus is in the cold room and the current kept low, natural convective cooling should be enough.

After the electrophoretic run, the block is sliced and the protein washed out from each slice with a little buffer. Resolution is not especially good and, as indicated above, would rarely be better than using ion exchange chromatography. On the other hand, the proteins remain in solution at all times and are subjected to less stress.

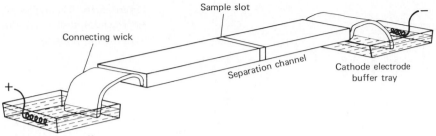

**Figure 5.14.** Horizontal preparative electrophoresis with powder or agar gel stabilization.

An interesting variant is isoelectric point electrophoresis (133) in which the isoelectric point of the enzyme is known and the mixture is prepared in a buffer at that pH. The sample is put into a slot prepared in agar gel, and electrophoresis carried out; the enzyme remains in the slot while contaminating proteins move away in both directions (Figure 5.15). The initial requirement is to find the pH at which the enzyme does not move. This might not be the same as its isoelectric point as determined by isoelectric focusing (cf. section 5.3). The isoelectric point may depend on the buffer used because of binding of buffer ions. Binding of phosphate or citrate ions is a common occurrence, and this lowers the isoelectric point. Isoelectric focusing approximates a zero buffer concentration, and the values found are consequently "isoionic points," with no buffer ions bound to the protein molecule.

Another factor affecting the real mobility, as opposed to the theoretical one, is electroendosmotic flow. This is the flow of bulk liquid solvents under the influence of an electric field due to immobilized charge groups on the walls of the electrophoresis channel and on any stabilizing powder or gel used.

Each ion has associated with it a number of water molecules which migrate with the ion. Immobilized anions such as carboxylates on starch cannot migrate, but the counterions associated with them do; being cations they move toward the cathode, resulting in a net movement of water in this direction. (The net result of water being carried by other buffer ions, in both directions, is just about zero.) With materials such as starch and agar, electroendosmosis can be

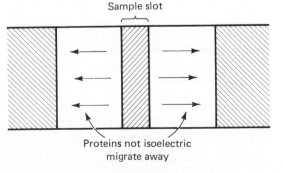

**Figure 5.15.** Isoelectric point electrophoresis; the enzyme required does not move, while other proteins with different isoelectric points migrate away in each direction.

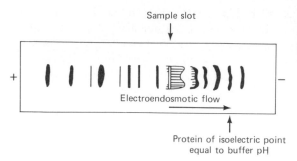

Sample slot

Electroendosmotic flow

Protein of isoelectric point
equal to buffer pH

**Figure 5.16.** The effect of electroendosmosis on the migration of proteins.

so large that many components are swept toward the cathode despite having a negative charge (Figure 5.16).

Liquid flow towards the cathode occurs as a result of immobilized negative charges (carboxylate, sulfate, phosphate) which are common on biological polymers. Even the most purified agarose has a few negative charges still on it, and electroendomosis can only be eliminated by gelling in the presence of a small quantity of positively charged polymer such as DEAE-dextran to neutralize the effect. In preparative electrophoresis the electroendomosis is not a serious problem, since the *cause* of the movement of a protein band is of less concern than the separation achieved from other protein components.

## Vertical

Horizontal "block" electrophoresis has not had a widespread application, although it does not require complicated apparatus. Vertical column electrophoresis has been used more extensively, especially in gels. Without a gel, there are two possible methods for stabilizing the liquid column, and these also apply to isoelectric focusing and isotachophoresis (cf. section 5.3). One is to use the same principle as in the horizontal system, with an inert powder filling much of the space, leaving the interstices as the effective volume in which protein migration occurs. The second is to stabilize gravitationally with a sucrose density gradient. A dense sucrose solution is placed at the bottom of the column, grading to zero or near zero at the top where the sample is applied. Note that the sucrose in the column is only to stabilize the liquid, minimizing convectional currents. It affects the proteins' migration somewhat by increasing viscosity as they move down, thereby slowing them, but at the same time diffusion is decreased. In a column, the normal procedure for removing the proteins after electrophoresis is simply to allow the liquid to run out, collecting the fractions with a fraction collector. This should be done slowly to avoid distorting the separated bands of protein as they move down the column.

## Vertical Gel Systems

The methods described above result in only crude separations; in free solution overlapping due to diffusional mixing is extensive, and the best results are rarely better than can be achieved by other methods. Gel (small-pore) methods are greatly preferable because of the high degree of resolution. However, there is one complication, which results in a necessarily much more complex apparatus, and that is the problem of getting the proteins out of the gel after electrophoresis is complete. Simply applying hydrostatic pressure will not do; surface tension pressure is far too great. Chopping up the gel is possible, but even after finely mincing the particles, recovery of (active) protein is usually very low. Consequently the procedure adopted is to allow each component to migrate right out of the gel column and collect it as it emerges into a perpendicular flow of buffer. This principle is adopted on the LKB Uniphor, a commercial apparatus sold for preparative electrophoretic techniques (Figure 5.17). Unfortunately, the bands, quite concentrated in the gel, become extensively diluted as they are swept away by the buffer, and this could be a disadvantage if the enzyme desired is unstable in dilute solution. A continuous-flow hollow fiber ultrafiltration concentrator can be used in conjunction with this apparatus so that the fractions quickly become more concentrated again before collection. Because of the variety of possible methods and home-made apparatuses, there are a number of different techniques described in the literature for preparative electrophoresis techniques, and the beginner is advised to consult these before deciding what system suits his particular needs and budget (134–138).

## Buffer Systems

The composition of the buffer used for pH control in electrophoresis is very important, because most of the current flowing through the separation channel is carried by the buffer ions. Very mobile ions (e.g., metal ions $Na^+$, $K^+$, $Mg^{2+}$, and simple anions $F^-$, $Cl^-$, $Br^-$, $SO_4^{2-}$, $HPO_4^{2-}$) carry much current because of their high mobility. This leads to heating, and unless there is a particular reason for the presence of such ions, they should be avoided. Bulky organic ions

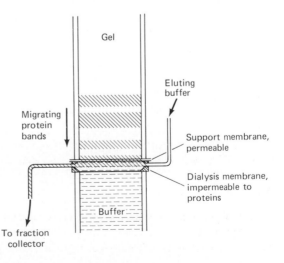

Gel

Eluting buffer

Migrating protein bands

Support membrane, permeable

Dialysis membrane, impermeable to proteins

Buffer

**Figure 5.17.** Elution of emerging protein bands from a vertical gel column as electrophoresis proceeds.

To fraction collector

have much lower mobility, especially if they are buffering species of the type $n = +1$ or $n = -1$ (cf. section 6.1). On the other hand, it is likely that small, simple cations would be more effective in preventing protein–protein interactions than these bulky ions, so if this is a problem, some simple salts may be needed.

Two systems of buffers for simple electrophoresis will be described—discontinuous and continuous systems. The latter are simple; the same buffer is used throughout so that in the electrophoresis channel no change in buffer composition occurs. In order to lessen voltage losses across the connections from electrodes to the actual electrophoresis channel, a buffer of higher concentration may be used (Figure 5.18a). Normally buffer concentrations with total ionic strength in the range 0.05–0.15 are used. This is a compromise between low conductivity, which allows application of a higher voltage but may result in protein-protein interactions, and a minimum interaction system with higher conductivity where heating may be a problem unless the current is kept down. So many different buffer systems have been used that it would be unreasonable to list them here. For simple gel electrophoresis a nondiscontinuous buffer widely used is Tris borate at around pH 8, or for somewhat higher pH values, 2-amino-2-methyl-1,3-propanediol glycinate, pH 9–9.5. The ionic strength of these buffer mixtures depends on the pH chosen; up to 0.2 M of the base may be required at the higher pH values because most of it will remain uncharged, and so not contribute to conductivity. High pH values are usually chosen so that the majority of proteins migrate toward the anode. But for special purposes many low-pH buffer systems have been described.

Discontinuous buffer systems are used in gel electrophoresis to sharpen up protein bands. Mainly seen in analytical systems, they can also be used in preparative electrophoresis. The principle is that a Kohlrausch discontinuity is

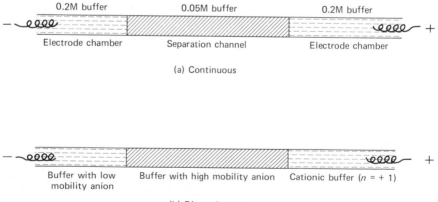

**Figure 5.18.** Buffer arrangements in (a) continuous and (b) discontinuous buffer system.

established between the starting buffer ions and the following ions which have a lower mobility (Figure 5.18b). The discontinuity junction is self-sharpening, and can often be observed as a result of refractive index changes due to sharp concentration changes at the discontinuity junction. The features which cause protein band sharpening are (i) a low conductivity immediately behind the junction, which implies a local high potential gradient—thus trailing protein molecules migrate more rapidly to catch up with the main band in front; and (ii) a higher pH immediately behind the junction—this has the same results as (i).

Typical discontinuous buffer systems (with the discontinuity being in the *anions*) involve phosphate, citrate, EDTAate, or chloride as the mobile starting anion, to be followed by a relatively slow-moving anion which is usually a buffer of the type $BH = B^- + H^+$. The low mobility of the follow-up buffer is partly due to the fact that a proportion of it is uncharged at a given instant. At pH 8–10, borate ions or glycinate ions have most often been employed. For pH values in the range 6–8, zwitterionic ions such as morpholinoethane sulfonate (Mes) or morpholinopropane sulfonate (Mops) are successful (139).

## Affinity Electrophoresis

As with other affinity techniques, the principle of affinity electrophoresis is to alter the behavior of a particular enzyme by introducing a ligand specific to that enzyme. In this case the behavior is the electrophoretic mobility. There are two possible modes of operation. First, inclusion of the (charged) ligand in the buffer should result in an increased or decreased mobility, depending on the difference between net charge with and without ligand. Preparative electrophoretic separation in the absence of ligand selects out a batch of proteins with the same average mobility. Then re-electrophoresis of this sample in the presence of the ligand should result in a separation of the enzyme required from the nonspecific proteins (Figure 5.19). This is so far little more than a theoretical idea.

The second procedure, which has been used successfully, involves immobilizing or trapping the ligand in the electrophoresis medium, such as on the gel or stabilizing powder used in the electrophoresis channel. This method is analogous to the widely used analytical procedures of rocket immunoelectrophoresis. It has been used in a number of ways (140–142), including a preparative method where the immobilized ligand retards the movement of the enzyme in the electrophoresis channel, and a method of investigating and determining interaction constants. The main successes have involved immobilized lectins for separating glycoproteins (141) and immobilized Cibacron Blue (cf. section 4.6) for retarding dehydrogenases (142). A recent review on the subject has been presented (143) which describes the problems and possible future developments of the technique.

## 5.3    Isoelectric Focusing and Isotachophoresis

Simple electrophoresis involves separation of proteins on the basis of their mobility at a particular pH. Isoelectric focusing involves setting up a pH gradient and allowing the proteins to migrate in an electric field to the point in the system where the pH equals their isoelectric point (Figure 5.20). To establish a pH gradient involves the use of polymeric buffer compounds which rather resemble proteins themselves, as they have large numbers of both positive and negative charges and possess isoelectric points in the same pH range. Such buffers are called ampholytes; they too migrate to their isoelectric point, and as there are hundreds or even thousands of individual ampholyte species, they spread across the whole slab/column between the cathode contact (a weakly conducting base) and the anode contact (a weak acid). Thus, after a few hours of application of the electric field the ampholytes have migrated and formed a pH gradient; the pH range depends on their composition (i.e., the range of isoelectric points of the ampholytes themselves). Any proteins present also move; since they are mostly larger, they take a little longer to reach the isoelectric position.

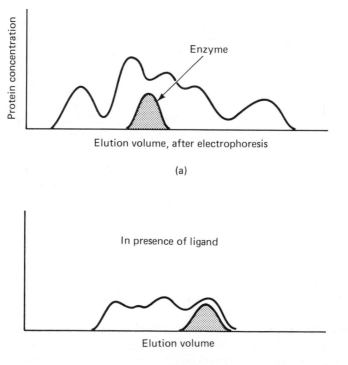

**Figure 5.19.** Principle of first type of affinity electrophoresis (see text): (a) with no ligand, (b) fraction containing enzyme re-run with added ligand.

**Figure 5.20.** Principles of isoelectric focusing.

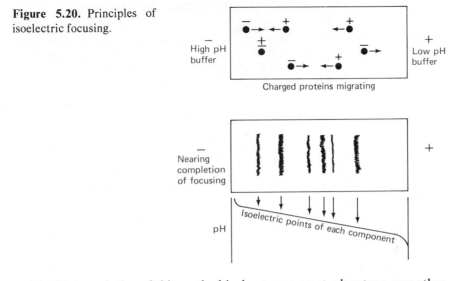

The high resolution of this method is due to one great advantage over other forms of electrophoresis. This is implied in the term *focusing*. In other protein solution methods (with the exception of isotachophoresis—see below) diffusion and zone mixing increase with time. In isoelectric focusing diffusion is countered, because as soon as a protein molecule diffuses away from its isoelectric zone, it becomes charged and so migrates back again. When focusing is complete, theoretically nothing moves in the system, so there can be no current (there are no other ions present). Consequently high electric fields can be applied with little heating, and the focusing is rapid. In practice a small current flows; if the applied field is too great, the gradient can break down and become less clearly defined.

Isoelectric focusing (electrofocusing) was originally designed as a preparative method; however, it has found more widespread use as an analytical procedure (cf. Chapter 9). Although an excellent procedure in theory, and also often in practice, several problems limit its usefulness. First is apparatus design. Isoelectric focusing is more flexible in its demands than conventional electrophoresis because of a relative lack of heating; a number of novel systems have been proposed and occasionally used successfully. These include placing the sample plus ampholytes in a long, flexible tube and winding that tube into a horizontal coil. After separation, individual components should have separated gravitationally to the lower parts of the coil, which is cut and the material from each segment collected separately (Figure 5.21). Electrofocusing can also be carried out in a horizontal slab with a liquid stabilizer such as Sephadex powder as is used in horizontal slab electrophoresis. However, the concentrated protein bands are gravitationally unstable. They are more dense than the adjacent liquid. Thus, the proteins tend to collect in the bottom and disturb the ampholytic distribution. A simple free-solution apparatus which exploits this

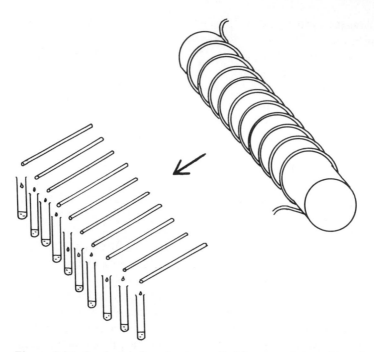

**Figure 5.21.** Isoelectric focusing in a coil. After completion of focusing the coil is cut and individual components recovered.

gravitational effect would be to use a corrugated base plate (Figure 5.22). However, these methods have not been developed extensively. The main procedure for preparative isoelectric focusing uses the vertical column system in an apparatus designed for ordinary electrophoresis. Stabilization of the column is commonly done using a sucrose gradient, although the other methods can be employed. But small-pore gels are not possible, since the samples cannot be removed from the gel; they do not migrate out of the gel column, but stay put at their isoelectric point. A very open gel such as agarose can be used as a stabilizing medium, if the gel can be removed from the column, sliced into

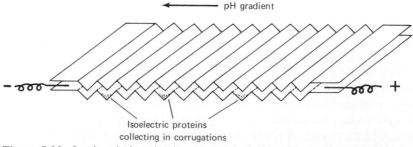

Isoelectric proteins
collecting in corrugations

**Figure 5.22.** Isoelectric focusing in a corrugated plate. [From Scopes (29).]

discs, and the proteins washed out. It is even possible to melt soft agarose gels, which is useful if the proteins are stable at about $45\,°C$.

Assuming a suitable apparatus, the next problem concerns the properties of the desired enzyme at its isoelectric point and, to a lesser degree, those of contaminant proteins in the preparation. Two features are essential: (i) the enzyme must be stable and (ii) it must be soluble at zero ionic strength. Neither of these factors can be taken for granted. Most intracellular enzymes have a sharp acid-instability curve in a pH range which may well overlap with their isoelectric points. Thus an animal protein may not survive pH 6.0 for long. Yet if its isoelectric point is 5.5 (by no means a low value), the method of isoelectric focusing demands exposure to a pH below 6.0. (Similar arguments arise for chromatofocusing, an alternative isoelectric focusing method; cf. section 4.3.) Plant and bacterial proteins may naturally exist at lower pH values (and so be more acid-stable) but also tend to have lower isoelectric points than animal proteins, so instability at the isoelectric point is equally likely.

If the stability conditions are suitable, then the solubility situation must be considered. In a column the formation of a precipitate at the isoelectric point will upset the electrical conductivity, and the precipitate may simply fall down to the bottom! However, apparatus of the type shown in Figure 5.22 might actually operate better if the protein precipitates at its isoelectric point. Note that the protein is unlikely to precipitate *close to* its isoelectric point; precipitation is a solubility feature, and high enough concentrations to cause precipitation are only likely to be present at the isoelectric point itself. Of course, if an impurity precipitates, it is likely to have an unsettling effect on the whole system.

A main reason why preparative isoelectric focusing has not been used widely is not theoretical, but strictly a matter of cost. Ampholytes supplied by LKB Products (Ampholines) or Pharmacia (Pharmalites) are not cheap, and their frequent use during development of an enzyme purification method can be a strain on the budget. At up to $400 per liter of mixed buffers, repeated large-scale trials are not economical. A typical column, suitable for applying 100–200 mg protein, might be 100 cm$^3$, with the ampholyte used at a concentration of 2–4% w/v. This is the total amount of ampholyte required, since the buffers between the separation column and the electrodes do not require any ampholyte. At the anodal end, a weak acid of the type $HA = H^+ + A^-$ is used, since any diffusion of this acid into the column causes it to ionize as it meets a higher pH, and the anions then return under electrophoresis. The protons are picked up by an ampholyte, which becomes positively charged, moves away from the acid end, and quickly neutralizes again, passing the proton on to the next ampholyte. In this way a small current passes, protons flowing from the anode to cathode. In an exactly analogous way, hydroxyl ions can flow from the cathode to the anode; the relative contribution of these two ionic flows to the total steady-state current depends on the pH range of the ampholytes and the $pK_a$ values of the acid and base at each end. Ampholytes covering large or small pH ranges are available, and if the isoelectric point of the enzyme is known with reasonable accuracy, a narrow range (e.g., 2 pH units) can be tried at once.

Because of the focusing, the sample can be applied at any position, or even mixed in with the ampholyte buffer when preparing the column. However, the latter is not recommended, since contact with quite low and high pH values at the ends may cause denaturation and precipitation. The top of the column is usually made the cathodic end, and the sample can conveniently be applied near the top (alkaline denaturation and precipitation are less likely than acid) and overlayed with a little more ampholyte before placing the cathode buffer on top. Care should be taken that the successive densities are correct; the sample protein will increase the density, so less sucrose should be present (Figure 5.23). Even if the stablizing system consists of powder, some consideration of the respective densities should be made; adding sucrose will not harm the system unless there are sucrose-metabolizing enzymes in the mixture. As with simple electrophoresis, a variety of different apparatuses have been described, each having its particular characteristics and operating schemes.

There is little doubt that isoelectric focusing, when used on protein mixtures which satisfy the criteria of stability and solubility at their isoelectric points, is the highest-resolution system, even though it does not discriminate between proteins on the basis of their molecular size. So many minor bands have been found with supposedly pure proteins that isoelectric focusing is the ultimate analytical test. However, there are instances where interaction between proteins and ampholytes of very similar isoelectric points results in multiple banding as an artifact. Because ampholytes are not present as a continuum but as discrete compounds with closely spaced isoelectric points, they too focus into bands (144). If an ampholyte of isoelectric point 6.1 interacts strongly with a protein of IEP 6.0, a conjugate of intermediate IEP is formed. Because of the

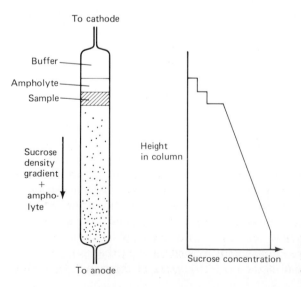

**Figure 5.23.** An isoelectric focusing column with sucrose density stabilization.

low charge on either component at around pH 6.0, the electric force is not sufficient to separate them. Thus a protein of IEP 6.0 might show other conjugate bands around that value which are not due to heterogeneity of protein, but rather to interaction with discrete molecules of ampholyte. This possibility is decreased by having ampholytes with more individual components, in which case conjugates would have IEP values so close together that discrete bands would not be resolved; the only effect of conjugate formation would be a diffusion of the protein position.

Considerable work has been done on the use of low-molecular-weight, low-conductivity buffers for establishing a pH gradient as an alternative to the expensive ampholytes (145–147). Although successful separations in such systems have been presented, the resolution, because of higher conductivity, has never been as impressive as when using polymeric ampholytes.

*Isotachophoresis* is the third method of separating proteins in an electric field. Simple electrophoresis is in a constant electric field, at constant pH, and separates according to actual mobility. Isoelectric focusing uses a constant electric field but a pH gradient, allowing the components to move until they become uncharged. Isotachophoresis also develops a pH gradient, and allows all the components to move at the same speed through an electric field that is of different strength at the location of each component; whereas it separates on the basis of mobility (mobility per unit electric field), *actual* velocities of each component are identical. This does not involve a complex monitoring of voltage gradients in order to apply different electric fields at various points; the system is self-forming (just as the pH gradient in isoelectric focusing is self-forming), provided that there is no excess of current-carrying ions to disrupt the boundaries between components. In the absence of ions of intermediate mobility, two like-charged ions of different mobility separate until all the faster ones are ahead of the slower ones, at which point a sharp boundary forms. The faster ions cannot actually move ahead of the slow ones and leave a void, since there are counterions traveling in the opposite direction across the boundary whose charge must be balanced at each location. Consequently a sharp discontinuity develops, known as a Kohlrausch discontinuity (Figure 5.24).

When the ions in question are proteins of different mobilities, they order themselves according to their mobilities just as do other ionic components. Generally the pH of the system is high, so that most proteins are negatively charged and move toward the anode in the applied electric field. Ideally the proteins are first adjusted to a pH using a cationic buffer, e.g., Tris, then dialysed against dilute buffer which contains a fast anion such as chloride. At first, current is carried by the chloride ions, which run ahead of the proteins; the only anions left behind are the proteins, which order themselves according to their mobility. Current is also carried in the opposite direction by $HTris^+$ ions; these are also present in the anode buffer so that they are constantly replenished. The anionic component of the anode buffer is not important, although it is worth remembering that chloride is not a good anion at the anode because the net electrolysis product is chlorine. Anions at the anode should preferably

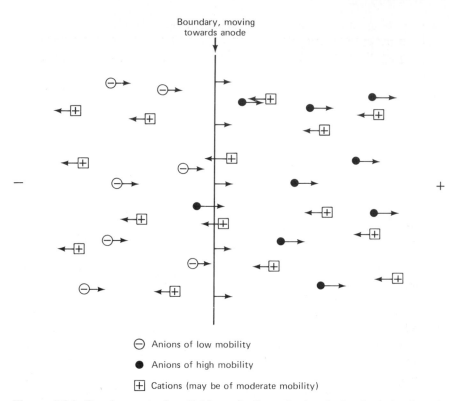

Figure 5.24. Development of a Kohlrausch discontinuity during isotachophoresis (also see Figure 5.18b).

liberate only oxygen, e.g., sulfate, borate, or phosphate. But at the cathode the anions will migrate toward the proteins, and should be chosen so that they have very low mobility. $\epsilon$-Amino caproate is used as a trailing anion in the buffer; LKB recommend 0.23 M $\epsilon$-amino caproate, pH to 8.9 with Tris; at this pH only about 2% of the $\epsilon$-amino caproate has a net charge [$NH_2-(CH_2)_5-COO^-$), the remainder being zwitterionic and so uncharged. Not only does the ion have a low mobility in itself, but the effective mobility is the average of this plus that of the uncharged form, i.e., 50$\times$ less. So this anion does not catch up any but the lowest mobility protein.

Because the bands of each component are constrained to move at the same velocity as one another, but have different mobilities ($u_i$), the voltage gradient ($E_i$) across each component self-adjusts so that actual velocity ($E_i \times u_i$) is the same for each component. This leads to a complex voltage variation along an isotachophoretic separation column, as shown in Figure 5.25. At each boundary there is a sudden drop in voltage and, because of the effects on protons, a transient pH change also. In theory, isotachophoresis could operate at constant pH if there were a strong enough buffer present. In practice, a pH gradient

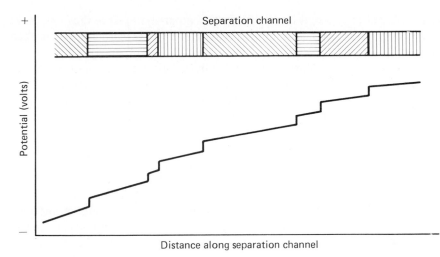

**Figure 5.25.** Variation of voltage along isotachophoresis channel during migration of the components.

becomes established; the more mobile proteins at the starting pH move into a lower pH region and become less mobile. Thus two effects cause the self-adjustment of mobilities: a differential voltage gradient, and a pH gradient operating to decrease the mobilities of the most mobile components. Which of these is the more important depends on the conditions of operation. With pure proteins alone the voltage gradient effect would probably be more important.

The conductivities of protein molecules are so low that a stable boundary can only be developed at extremely (and dangerously) high potentials. To overcome this, preparative isotachophoresis of proteins is carried out in the presence of ampholytes, which act as spacers, having mobilities intermediate between the protein components, increasing the conductivity, and helping to form more stable boundaries. Their presence helps a pH gradient to develop. Thus isotachophoresis of proteins resembles isoelectric focusing, except that the components continue to move (and all eventually pass out of the column), and the individual components never reach the pH value of their isoelectric point. In this way three disadvantages of isoelectric focusing are overcome; the proteins move out of the separation channel, and can be collected on the same principle as for simple electrophoresis. Instability at the isoelectric point need not be a problem if the pH remains above that value (or below, if the whole direction of flow is reversed), and insolubility at the isoelectric point need not be of any concern. The main disadvantages in comparison with isoelectric focusing are poorer resolution, especially as the theoretical sharpness between zones can never be maintained right through to the fraction collector, and the fact that interspaced ampholytes do not completely separate adjacent proteins. But another advantage over both simple electrophoresis and isoelectric focusing is that of high capacity; quite large amounts of protein can be processed (148).

Also, it is not as wasteful in consumption of ampholytes, since they only have to be added to the protein sample, perhaps 2–5 times more ampholyte than protein, by weight.

Because of the complexities of the apparatus and relative ignorance of the method, few really successful applications of preparative isotachophoresis have been reported in the literature. But the method has considerable promise and no doubt in the future it will be used more extensively.

## 5.4   Liquid Phase Partitioning

This section has been included for completeness; partitioning of proteins between different phases has not been a notably successful method for separation, though very little work has been done on it. The technique of countercurrent distribution, a popular method in organic chemistry, was applied earlier to certain proteins, but with limited success. The main problem was to find water-immiscible (or partially miscible) solvents that proteins were soluble and stable in; available solvents were appropriate only for very small hormone-type polypeptides. Two-phase systems suitable for proteins are now available (see below), which could be used in the countercurrent distribution mode.

Because few proteins are soluble, let alone stable in organic solvents, liquid phase partitioning a priori would not seem to be useful. However, addition of certain pairs of hydrophilic polymers to aqueous solutions can cause a phase separation without the presence of any hydrophobic solvent (149). In particular, dextran and polyethylene glycol when dissolved in water in appropriate proportions develop two phases, a dextran-rich one which is denser and settles on the bottom and a polyethylene glycol-rich phase on top. With more polymer components, still more phases can separate out; 18-phase systems have been described (150). In such circumstances a protein mixture distributes itself through the solution, and depending on a variety of solubility properties of the individual protein components, relative enrichment of a particular protein can occur in one of the phases. Separation of the phases is simple; a brief centrifugation may help to sharpen the boundary, after which a separatory funnel can be used. The partitioning of a particular component in this system is unlikely to be all-or-nothing, so recovery from a one-step procedure may not be good. The distribution of various proteins in a 3-phase PEG-dextran-Ficoll system has been reported (151). The pH of the system was found to be an important factor.

A much more effective development, which should see further use in the future, is the use of affinity partitioning. In this case one of the solutes has a ligand attached covalently to it; the phase in which this solute concentrates then attracts the bulk of the specific enzyme. Purification of steroid isomerase by coupling estradiol to polyethylene glycol and dissolving this plus dextran into a crude preparation resulted in a good separation from impurities (152).

Cibacron Blue F3GA and Procion Red HE-3B linked to polyethylene glycol have also been used successfully, particularly in the commercial-scale production of glucose 6-phosphate dehydrogenase (153). The technique is gentle and rapid; a good degree of purification can be achieved if most of the contaminants can be persuaded to partition into the phase without bound ligand. On the other hand, there are obvious problems such as the removal of the enzyme from soluble polymer–ligand complex, and the technique is unlikely often to be superior to affinity adsorption in resolution. But, because it does not involve adsorption to a solid phase, it is gentler, and so more applicable to labile proteins. In particular, it is suited to particulate (membrane-bound) fractions; turbid suspensions can be used.

Single-step partitioning is not satisfactory unless the distribution between the two phases is highly unequal. If the partitioning could be adopted in a chromatographic procedure, separation of components of even quite similar partition coefficients should theoretically be possible. Liquid-liquid chromatography does not at first sight seem to be possible. But in fact it is possible, with one of the liquid phases immobilized. The method has been successfully demonstrated for separation of nucleic acids (154), and there seems no reason why it should not be adopted for proteins. By prequilibration of a suitable support medium such as agarose beads with one of the phases (e.g., the dextran-rich phase), and running the other phase as the moving buffer in the column (Figure 5.26), surface tension prevents mixing of the two components; the dextran-rich phase remains within and immediately around the surface of the beads. Simple partitioning of proteins on such a column is possible, but affinity partitioning should be even more effective. The system becomes very analogous to conventional affinity chromatography in this case, the only difference being that the bound ligand is not immobilized covalently to a solid support, but trapped within the beads in a soluble form.

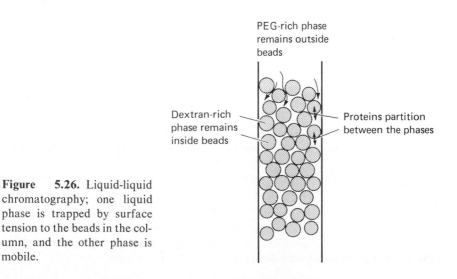

PEG-rich phase remains outside beads

Dextran-rich phase remains inside beads

Proteins partition between the phases

**Figure    5.26.** Liquid-liquid chromatography; one liquid phase is trapped by surface tension to the beads in the column, and the other phase is mobile.

## 5.5  Ultrafiltration

A final method to note in this chapter is the use of ultrafiltration, using gas pressure to force liquid through a membrane (cf. Figure 1.13b). If the membrane pores are such that smaller protein molecules can pass through with the "ultrafiltrate," a separation is achieved. Larger molecules are retained and concentrated relative to the starting solution. Because of the range of pore sizes in an ultrafiltration membrane, there is no absolute cutoff point. A proportion of

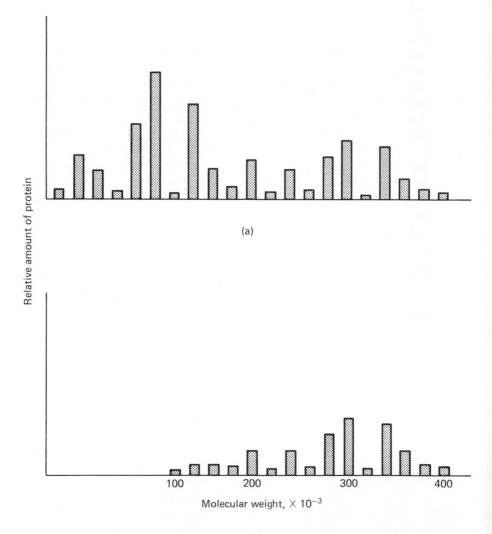

**Figure 5.27.** A possible distribution of proteins by size and amount before (a) and after (b) ultrafiltration through a membrane with pores of nominal cutoff at 200,000 daltons (proteins all assumed spherical).

molecules of sizes close to the stated cutoff size will pass through, the remainder staying behind. But if the protein of interest has a size much smaller or larger (say, 30–50% at least) than the quoted cutoff size, virtually all of it will appear in either the ultrafiltrate or the "retentate."

The method is obviously less discriminating than gel filtration (though based on the same principle, size of molecules), as it only provides two fractions, "bigger molecules" and "smaller molecules." Nevertheless, there are advantages in certain applications, especially when one has a large volume of dilute protein solution such as may be obtained from an adsorption column. Before gel filtration this would need concentrating, probably by ultrafiltration, so if a membrane is chosen that just allows retention of the protein required, a separation of the smaller proteins is achieved without any extra effort. This is of most benefit when purifying proteins of larger size. For instance, suppose the protein we are interested in has a molecular weight of 300,000 and other proteins are distributed in size and amount as illustrated in Figure 5.27a. Using a membrane with nominal cutoff of 200,000, most of the protein of size less than 150,000 passes through the membrane, and the resulting distribution is shown in Figure 5.27b. In this example, more than half the protein has been removed, which allows the retentate to be concentrated to a smaller volume. On the other hand, the result following a subsequent gel filtration will be just the same as if the ultrafiltration had not removed smaller proteins. But because less protein is present at the start, a smaller gel filtration column can be used.

Ultrafiltration slows up as the protein concentration increases, so redilution with buffer once or twice may be useful before finally collecting the retentate. This is not so desirable if the protein wanted is in the ultrafiltrate, since it is already dilute, and extra buffer passing through makes it more so. A second ultrafiltration using a membrane with smaller pores can be used to concentrate the first ultrafiltrate.

## General References

L. Fischer (1980), Gel filtration chromatography. In *Laboratory techniques in biochemistry and molecular biology* (T. S. Work and R. H. Burdon, eds.), Vol. I, Part II, rev. ed., Elsevier/North-Holland, Amsterdam.

A. H. Gordon (1979), Electrophoresis of proteins in polyacrylamide and starch gels. In *Laboratory techniques in biochemistry and molecular biology* (T. S. Work and R. H. Burdon, eds.), Vol. I, Part I, rev. ed., Elsevier/North Holland, Amsterdam.

Pharmacia Fine Chemicals AB Publications (1980), *Gel filtration: Theory and practice.* Uppsala.

# Chapter 6
# Maintenance of Active Enzymes

## 6.1   Control of pH: Buffers

A key requirement in enzyme work is to know the pH at all times; to control the pH a buffer is essential. It is also important to know the characteristics of the buffer being used and to have on hand a range of different buffers so that one most suited to the application is available. Over the decades of enzyme studies a great number of buffers have been used, and new ones are still being developed. There are several characteristics we need to know about a buffer:

(1)  $pK_a$ value
(2)  Variation of $pK_a$ with temperature, ionic strength
(3)  Anionic, cationic, or multiple charges on buffer species
(4)  Interaction with other components, e.g., metal ions
(5)  Solubility
(6)  Expense
(7)  UV absorption

A buffer maintains pH nearly constant by absorbing protons produced in other reactions, or releasing them if they are consumed. Optimum buffering is at the point where both these capabilities are equal. But for some purposes, e.g., when it is known that protons are to be released, the optimum buffering may be when the basic form is in excess—it depends on how much acidification is expected and how much pH change can be tolerated.

The simplest types of buffer are those in which one form is uncharged, i.e.,

$$HA = H^+ + A^-$$

or

$$HB^+ = H^+ + B$$

The dissociation constant is defined in the first case as

$$K_a = \frac{[H^+][A^-]}{[HA]}$$

or in the second case as

$$K_a = \frac{[H^+][B]}{[HB^+]}$$

and is usually expressed in logarithmic form:

$$pK_a = -\log K_a$$

Thus

$$pK_a = -\log \frac{[H^+][A^-]}{[HA]} = -\log[H^+] - \log \frac{[A^-]}{[HA]}$$

$$= pH - \log \frac{[A^-]}{[HA]}$$

This is the Henderson–Hasselbach equation, which when rearranged indicates the pH obtained with a solution containing both forms of the buffer:

$$pH = pK_a + \log \frac{[\text{basic form}]}{[\text{acidic form}]}$$

This equation applies to both types of buffer, HA and $HB^+$. A few simple calculations show that $pH = pK_a$ when [basic form] = [acidic form], $pH = pK_a + 1$ when [basic form] = $10 \times$ [acidic form], and $pH = pK_a - 1$ when [basic form] = $0.1 \times$ [acidic form]. Generally it is desirable to work within about 0.5 unit of the $pK_a$ because then (a) there is a reasonable amount of each form present, and (b) the pH changes less per proton released or adsorbed the closer one is to $pK_a$.

In the equations below, the following terms are used:

$z$ = charge on a given species
$n = 2z - 1$, where $z$ = charge on the acid buffer form

The $n$ value is useful in indicating the effect of ionic strength on $pK_a$ values (see below). By definition, $n$ is always an odd number.

$pK_a^0 = -\log$ (acid dissociation constant), extrapolated to zero ionic strength
$pK_m = pK_a$ for the $m$th dissociation starting from low pH (for multicharged buffers)
$I$ = ionic strength = $\frac{1}{2}\sum_i C_i(z_i)^2$, where $C_i$
      = concentration of a charged species and $z_i$ is its charge

For simplification it has been assumed that activity coefficients are unity; at higher concentrations this becomes a less valid assumption.

The concentration of buffer required is really a feature of the system itself rather than the buffer, although if one has to work at the edge of the useful pH range, e.g., around 1 pH unit from $pK_a$, more buffer is needed to maintain buffering power. It is useful to know what direction the most likely change in pH might be, so that the buffer can titrate in and counteract the change more strongly the more it occurs. For example, suppose we wish to have a solution at pH 6.5 with maximum buffering at a low ionic strength, resisting acidification. An imidazole buffer may be suitable, but as its $pK_a$ is 7.0, acidification from pH 6.5 downward would tend to weaken its power. But if histidine is used ($pK_a$ 6.0), acidification would titrate the buffer and make it stronger as the $pK_a$ was approached. Alternatively, it may be required that the buffer be anionic, in which case N-morpholinoethane sulfonate (Mes, $pK_a$ 6.2) would be most appropriate.

Many of the buffers which have been used in the past are no longer the best choice because of the ready availability of an alternative. These include maleate $pK_a$ 6.3 (usually in the form of Tris-maleate, which covers a big pH range though with a weakness around 7.0), because it has UV absorption and $n = -3$ (see below); barbitone $pK_a$ 8.0 (UV adsorption and poisonous—also known as veronal); glycylglycine $pK_a$ 8.2 (Tricine has marginally better characteristics and is cheaper); cacodylate $pK_a$ 6.2 (poisonous, expensive); and citric acid ($pK_2$ 4.8, $pK_3$ 6.4, binds to some proteins, complexes metals, and $n = -3$ or $-5$).

For acetic acid, the acidic buffering species is $CH_3COOH$, for which $z = 0$; $n = -1$ for acetic acid/acetate buffer. For citric acid $pK_3$, the acidic species is citrate$^{2-}$, $z = -2$, $n = -5$. For Tris, the acidic species is $HTris^+$, $z = +1$, $n = +1$.

## Effect of Temperature and Ionic Strength on $pK_a$ Values

The ionic strength is greater for a given strength of buffer if $|n|$ is greater than 1. More importantly, $n$ is an important indicator of the effect of ionic strength on the $pK_a$ value. The parameter $n$ appears in the simplified Debye-Hückel equation relating $pK_a$ with $pK_a^0$ ($I = 0$):

$$pK_a = pK_a^0 + \frac{0.51 n I^{1/2}}{1 + 1.6 I^{1/2}} \quad \text{(at 25°C)} \tag{6.1}$$

Thus values of $|n|$ greater than 1 result in larger effect of ionic strength on $pK_a$, especially so for citrate (Table 6.1), for which $n = -5$ in the pH range citrate is usually employed. Note that for cationic buffers such as Tris, $pK_a$ rises with ionic strength. The effect of ionic strength should be remembered when making up stock buffers of greater strength than to be used.

**Table 6.1.** Effect of Ionic Strength of $pK_a$ of Some Characteristic Buffers[a]

|            | $n$  | $pK_a^0$ | $pK_a$, $I = 0.01$ | $pK_a$, $I = 0.1$ | Concentration of buffering species at the $pK_a$, if $I = 0.1$, entirely due to buffer |
|------------|------|----------|--------------------|-------------------|-----------------------------------------------------------------------------------------|
| Acetate    | $-1$ | 4.76     | 4.72               | 4.66              | 0.2                                                                                     |
| Phosphate  | $-3$ | 7.20     | 7.08               | 6.89              | 0.05                                                                                    |
| Citrate    | $-5$ | 6.40     | 6.29               | 5.88              | 0.022                                                                                   |
| Tris       | $+1$ | 8.06     | 8.10               | 8.16              | 0.2                                                                                     |

[a]Theoretical, calculated from Eq. (6.1) at 25°C.

Temperature also affects $pK_a$ values, especially for buffers involving amines. The effect of temperature on the buffering can be judged from the thermodynamic expressions:

$$\frac{d \ln K_a}{dT} = \frac{\Delta H^0}{RT^2} \quad , \quad \frac{dpK_a}{dT} = -\frac{\Delta H^0}{2.3RT^2}$$

and

$$\Delta G^0 = \Delta H^0 - T\,\Delta S^0 = -RT \ln K_a = 2.3RT\,pK_a$$

Thus the value of $\Delta G^0$ is large and positive for high $pK_a$ buffers, and depending on the entropy of dissociation, $\Delta H^0$ likewise tends to be large and positive. Thus the rate of change of $pK_a$ with temperature is likely to be negative and larger the higher the $pK_a$ value. On the other hand, low-$pK_a$ buffers change less with temperature. The entropy of dissociation is very small for amine buffers which are widely used for pH values above 6, this being a result of no formation of new charged ions when an $n = +1$ buffer dissociates, e.g., Tris:

$$HTris^+ \rightarrow Tris + H^+ \qquad \Delta S^0 = 5.7 \text{ J deg}^{-1} \text{ mol}^{-1}$$

The entropy change depends mainly on the degree of participation of water molecules in hydrating the species; the small positive $\Delta S^0$ for Tris (Table 6.2) indicates that the protonated species is probably little more hydrated than the neutral form:

$$HTris^+ \cdot (H_2O)_n \rightarrow Tris \cdot (H_2O)_m + H_3O^+ + (n - m - 1)H_2O$$

where $n \geq (m - 1)$.

On the other hand, for an $n = -1$ buffer, two new ions are formed where none existed before, e.g., acetic acid:

$$HAc \rightarrow Ac^- + H^+ \qquad \Delta S^0 = -90.5 \text{ J deg}^{-1} \text{ mol}^{-1}$$

Assuming that the acetate ion is more hydrated than the neutral acid, several molecules of water become "more ordered," and so there is a large negative entropy change:

$$(n - m + 1)H_2O + HAc \cdot (H_2O)_m \rightarrow H_3O^+ + Ac^- \cdot (H_2O)_n$$

where $n > m$.

**Table 6.2.** Thermodynamic Values for a Range of Buffers[a]

| | $n$ | $pK_a^0$ | $dpK_a/dT$ (at 25°C) | $\Delta H^0$ (kJ mol$^{-1}$) | $\Delta G^0$ (kJ mol$^{-1}$) | $\Delta S^0$ (J deg$^{-1}$ mol$^{-1}$) |
|---|---|---|---|---|---|---|
| Phosphoric acid $pK_1$ | $-1$ | 2.15 | $+0.004$ | $-6.8$ | 12.3 | $-64$ |
| Acetic acid | $-1$ | 4.76 | $-0.0002$ | 0.3 | 27.3 | $-90$ |
| Citric acid $pK_3$ | $-5$ | 6.40 | 0 | 0 | 36.6 | $-123$ |
| Imidazole | $+1$ | 6.95 | $-0.020$ | 34.0 | 39.9 | $-19$ |
| Phosphoric acid $pK_2$ | $-3$ | 7.20 | $-0.0028$ | 4.8 | 41.3 | $-12$ |
| Tes | $-1$ | 7.50 | $-0.020$ | 34.0 | 43.0 | $-30$ |
| Tris | $+1$ | 8.06 | $-0.028$ | 48.0 | 46.2 | 6 |
| Ammonia | $+1$ | 9.25 | $-0.031$ | 53.0 | 53.1 | 0 |
| Carbonate | $-3$ | 10.33 | $-0.009$ | 15.4 | 59.3 | $-147$ |

[a]The values of $\Delta H^0$, $\Delta G^0$, $\Delta S^0$ were calculated from the $pK_a$ and $dpK_a/dT$, which were obtained from other sources (51,52).

Similar arguments can be made for $n = -3$ and $-5$ buffers, where the more charged the ion is, the more hydrated it becomes. Zwitterions which buffer in the higher pH range due to dissociation of a protonated amino group are slightly less affected by temperature than straight amines, for similar reasons: $n = -1$ rather than $+1$, although the neutral form is probably well hydrated as it has both negative and positive charges on it, leading to an intermediate $\Delta S^0$ value, e.g., Tes:

$$Tes^{\pm} \rightarrow Tes^- + H^+ \qquad \Delta S^0 = -30 \text{ J deg}^{-1} \text{ mol}^{-1}$$

In Figures 6.1 and 6.2 the $pK_a$'s of phosphate ($pK_2$) and Tris are plotted as a function of temperature and ionic strength. Note that the $pK_a$ of Tris changes

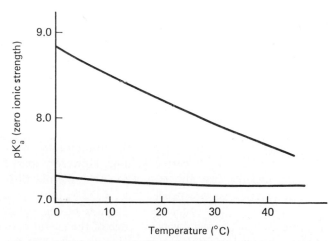

**Figure 6.1.** Variation of $pK_a$ with temperature for phosphate (lower line) and Tris (upper line). Note that temperature has only a small effect on phosphate, but a large effect on Tris.

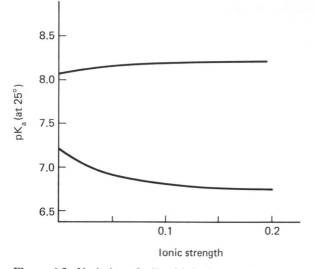

**Figure 6.2.** Variation of $pK_a$ with ionic strength for phosphate (lower line) and Tris (upper line). The effect on phosphate is greater by a factor of 3 (the ratio of the $n$ values), and is in the opposite direction (since $n$ is negative for phosphate and positive for Tris).

by 1 pH unit when going between the biochemist's traditional temperatures of 4 and 37°C.

Table 6.3 lists a series of useful buffers and indicates their properties. More extensive tables can be found (50,51). An ideal series of buffers would consist of two sets, one with $n = -1$, the other with $n = +1$. Suggested buffers for ion exchange chromatography are listed in Table 4.6. The only gaps that cannot be economically filled at present are:

(a)  An $n = -1$ buffer with $pK_a$ around 5.5 which has no UV absorption
(b)  An $n = +1$ buffer with $pK_a$ around 5.5 which is nonvolatile, with no UV absorption
(c)  An $n = +1$ buffer with $pK_a$ around 6.0 which does not complex metal ions

## Making up Buffer Solutions

Buffers can be made by weighing the appropriate amount of each form of buffer and making up to correct volume. However, apart from the dangers of making a mistake, it is always best to check the pH with a meter, especially when other additives are needed, e.g., EDTA, mercaptoethanol, magnesium chloride. Some buffers cannot be relied on to be completely anhydrous, so precise values of pH, especially at the edge of the useful buffering range, are not reliably obtained without checking. Ionic strength is not only important in affecting the $pK_a$ values, but may also be a vital variable in other methods such as ion exchange.

**Table 6.3.** A Selection of Useful Buffers

| Name | $n$ | $pK_a^0$ (25°C) | Comments |
|---|---|---|---|
| Lactic acid | −1 | 3.86 | |
| Acetic acid | −1 | 4.76 | |
| Pivalic acid (trimethylacetic acid) | −1 | 5.03 | Not very soluble; unpleasant odor |
| Pyridine | +1 | 5.23 | Volatile, poisonous |
| Picolinic acid | −1[a] | 5.4 | Strong UV absorption |
| Succinic acid | −3 | 5.64 | |
| Histidine | +1[a] | 6.0 | Complexes $Me^{2+}$ strongly |
| N-Morpholinoethane sulfonic acid (Mes) | −1[a] | 6.15 | |
| Bis-(2-hydroxyethyl)imino-tris-(hydroxymethyl)methane (Bis-Tris) | +1 | 6.5 | |
| N-(2-acetamido)-2-aminoethane sulfonic acid (Aces) | −1[a] | 6.9 | |
| Imidazole | +1 | 6.95 | Complexes $Me^{2+}$ |
| Phosphate | −3 | 7.20 | Stabilizes many enzymes |
| N-Morpholinopropane sulfonic acid (Mops) | −1[a] | 7.2 | |
| N-Tris(hydroxymethyl)methyl-2-aminoethane sulfonic acid (Tes) | −1[a] | 7.5 | |
| Triethanolamine | +1 | 7.75 | |
| Tris(hydroxymethyl)aminomethane (Tris) | +1 | 8.06 | |
| N-Tris(hydroxymethyl)methyl-glycine (Tricine) | −1[a] | 8.15 | |
| Tris(hydroxymethyl)aminopropane sulfonic acid (Taps) | −1[a] | 8.4 | |
| 2-Amino-2-methyl-1,3-propanediol | +1 | 8.8 | |
| Diethanolamine | +1 | 8.9 | |
| Taurine | −1[a] | 9.1 | |
| Ammonia | +1 | 9.25 | Volatile |
| Boric acid | −1 | 9.23 | Complexes with many carbohydrates |
| Ethanolamine | +1 | 9.5 | |
| Glycine | −1[a] | 9.8 | |
| 1-Aminopropan-3-ol | +1 | 9.95 | |
| Carbonate | −3 | 10.3 | |

[a]Indicates zwitterionic buffer in either acidic or basic form.

The ionic strength of a buffer may depend on how it is made up. For instance, "20 mM triethanolamine buffer, pH 7.5" might be 20 mM of triethanolamine adjusted to pH 7.5 with HCl, giving an ionic strength of 0.012 (acid form/basic form = 12:8 at pH 7.5); the only ions are H-triethanolamine$^+$ and Cl$^-$. Alternatively, it may be made by adjusting 20 mM triethanolamine chloride with NaOH to pH 7.5, in which case $I = 0.020$, since 8 mM NaCl would also be present. It might even be that triethanolamine was adjusted with sulfuric acid, in which case the doubly charged sulfate ions contribute more to the ionic strength, in this case $I = 0.018$. Authors rarely give full details of how they made up their buffers (including temperature); fortunately, it is not *always* critical. Concentrations of buffer used are fairly arbitrary; no one would suggest that because the buffer was 20 mM, 21 mM would not work. Don't waste time weighing out buffer salts to 5 or 6 places. The variable water content alone makes such action irrelevant. Unless otherwise stated, pH values are normally adjusted down with HCl or up with either NaOH or KOH for making up a buffer solution. To maximize buffering power for a given ionic strength the following should be observed:

(1) Choose a buffer with $pK_a$ within 0.5 unit of the desired pH, with $n$ value of $+1$ or $-1$.
(2) Use the uncharged form of the buffer, and adjust it up ($n = -1$) or down ($n = +1$).
(3) For absolute maximum buffering power, choose two buffers, one of $n = -1$ with $pK_a$ 0.5–1 unit higher than the other, of $n = +1$. Mix the two uncharged forms until the appropriate pH value is reached.

In example 3 it may be difficult to determine the ionic strength, especially if there are strong interactions between the charged forms of the buffers, but the conductivity is normally very low. Such buffers have been used widely for electrophoresis (cf. section 5.2), e.g., Tris-borate pH 8–9, Imidazole-Mops pH 6.8–7.5, since the ionic strength is low. All ions are buffers, and also the conductivity of the bulky ions is low.

It is useful to have a series of stock buffers 10 times, or even 100 times more concentrated than intended for use. The advantages are: the storage volume is small; bactericidal agents added (e.g., azide) are diluted out when the buffer is used; and the buffer can be added to an otherwise buffer-free protein solution without exposing the latter to extremes of pH. But remember that on dilution the pH will change (Figure 6.2, Table 6.1).

There will be frequent occasions when it is necessary to adjust the pH of a protein solution. Consider two situations. The first is a slight adjustment, to a pH which is within the buffering range of the buffer salts already present. In this case a dilute weak acid (e.g., acetic acid, not more than 1 M) or base (Tris, 1 M) can be used; the solution should be well stirred so that localized acid or alkaline areas around the drops of adjusting solution are quickly dispersed. Only if the pH values concerned are below 5 or above 8 might some stronger acid or base be needed. If a major adjustment in pH is needed and there is a strong buffer in the solution, more concentrated acid (or base) can be used to lessen the volume increase. So that localized extremes of pH are avoided, the adjusting solution itself can be partly neutralized. Using 5 M acetic acid pH 4.5 (NaOH) as a concentrated pH-lowering solution, no part of the enzyme mixture is ever exposed to a pH lower than 4.5. For raising pH values, 5 M ammonia pH 9.5 (HCl) can be used.

The second situation is where the final pH is outside the buffering range of the salts initially present. One could add suitable buffers before adjusting the pH, but it is generally better to use an adjusting solution that itself will form a buffer when the pH reaches the desired value. For example, suppose a solution is in Tris-chloride pH 8, and one wants to change the pH to 6.5. By using the acid form of a buffer with $pK_a$ in the range 6.2–6.8, the pH is lowered and the solution remains buffered. In this example one could use M Mes (acid), or if one wanted the buffer to be of the $n = +1$ type, M histidine-chloride might be appropriate.

The temperature at the time of making up a buffer is very important if precision in pH is necessary. Often methods for enzyme purification state "all procedures were carried out at 4°C", yet the buffers used were almost certainly mixed at room temperature, so the actual pH, at 4°C, would have been different. Most pH meters have a temperature adjustment knob, but this does not mean that by dialing up 4°C, and using the meter at 25°C, the correct compensation has been made—how could the meter know what value of $dpK_a/dT$ is appropriate for that solution?! To make up a buffer at 4°C (the usual cold-room temperature), set the temperature compensation knob at 4°C, place the electrode in standard buffer *at 4°C,* allow it to chill to that temperature, and set the reading to the standard value *at 4°C.* Then adjust the buffer, protein solution or whatever *at 4°C,* to the desired pH. In some methods based on isoelectric precipitation in the presence of cold organic solvent, the pH must be adjusted at +2°C, and the pH value must be correct to ±0.1 unit. The only way to do this reliably is as above. Incidentally, the pH changes in the presence of organic solvent—yet another factor affecting $pK_a$ values (cf. section 3.4).

This section has treated buffers assuming that pH can be measured correctly and accurately each time. Modern pH meters are easy to use and give a sense of confidence of accuracy, especially with a digital readout. Nevertheless, there are pitfalls in using a pH meter, especially with the common combined glass electrode. Even if the instructions are meticulously followed, incorrect readings can be obtained because of poor conductivity and the development of liquid junction potentials in the ceramic plug between the sample being measured and the KCl inside. Illingworth (155) recently investigated this phenomenon and found that many well-used and some new combined electrodes were giving pH values up to 0.4 unit wrong with buffers that had ionic strengths significantly different from that of the buffer used to standardize the pH meter. It is advisable to check your own meter. Standardize with a 50 mM phosphate buffer, then dilute this buffer 10-fold; the pH should read 0.2 unit higher (see Figure 6.2). If not, it would be worth attending to the ceramic plug, *gently* scraping across its surface to clear any clogging. It is best to keep the electrode in a strong KCl solution between measurements.

In conclusion, it is always vital to be sure of the pH. If a method does not state the pH values, then measure them yourself; variable success in applying the method may be due to pH variations in the raw material. If the buffer used in the method is unavailable, consider whether that buffer might have had some special influence, or whether it could be replaced with another of similar characteristics.

## 6.2   Stabilizing Factors for Enzymes

Ever since biochemists started purifying enzymes they have been plagued by the problems of loss of activity. The general reason is that enzymes are often sensitive protein molecules which when released from their natural protective environment are subjected to a variety of stresses they may not be able to resist. Extracellular enzymes are an exception simply because they are designed to operate in a more hostile environment. It is for this reason that many of the early successes in purifying proteins and determining structures were with extracellular enzymes, which tend to be small, tough protein molecules, often present as a relatively high percentage of the protein in the fluid concerned. But the natural environment inside a cell is very different. Soluble proteins, whether cytoplasmic or inside organelles, are present in very concentrated soups of protein, usually at least 100 mg ml$^{-1}$, and estimated as high as 400 mg ml$^{-1}$ in the mitochondrial matrix (24). There tends to be a relatively low (in some cases zero) oxygen tension, and a variety of reducing compounds are present, such as glutathione, to maintain a high reducing potential. Metabolites are also present in quite substantial amounts, and their presence may stabilize the enzymes. As soon as the tissue is disrupted, the proteins are released from this protective environment, diluted out into an extraction buffer, which is probably fully oxygenated, and proteolytic enzymes which were held in separate lysosomal compartments inside the cell are released also. The enzymes are exposed to three types of threat which may lead to loss of activity:

(1)  Denaturation
(2)  Inactivation of catalytic site
(3)  Proteolysis

Under type 2 may be included such things as loss of cofactors and dissociable prosthetic groups as well as chemical changes to the amino acid residues in the active site.

### Denaturation

Denaturation can be minimized if the principle effectors of denaturation are avoided. These are extremes of pH, temperature, and denaturants such as organic solvents (cf. section 3.5). The natural pH inside a cell would normally be in the range 6–8, so buffers within this pH range, or as close as possible to the actual pH in the tissue concerned, should protect against pH denaturations. Note that the physiological pH in most animal cells is 7.0–7.5 *at 37°C*. Because of the effects of temperature on the buffering components present, the equivalent physiological pH near 0°C for animal tissues is closer to 8.0 than to 7.0. Some problems encountered in controlling the pH when making the extract have been discussed earlier (cf. section 2.3).

It always used to be taken for granted that enzyme purification had to be carried out in the cold because enzymes are so labile; reducing the temperature by 20°C decreases the rate of most processes by a factor of 3–5, so coldness should be beneficial. However, it also slows up the processes involved in separation, e.g., in ion exchange chromatography, so if one has to increase the time factor by the same proportion, nothing is gained. This does not apply to denaturation, because of its extremely steep response to temperature (cf. section 3.5). But most proteins, especially from warm-blooded mammals, show little signs of heat denaturation below 40°C if the pH is near the physiological value—after all, they survived those conditions for perhaps days in situ. Only if dealing with species whose tissues are always cold, e.g., cold-water fish, may room temperature cause direct heat denaturation. Some enzymes even denature *because of the cold,* a phenomenon known as cold denaturation, due to weakening of hydrophobic forces holding the molecule together. So although it is generally advisable to keep solutions on ice while they are not being subjected to a fractionation process, it is sometimes better to warm them up when carrying out the fractionation. It is often not *necessary* to actually work in a cold room; it is always more comfortable not to do so.

Selective denaturation has been discussed (cf. section 3.5), as has the use of organic solvents as a fractionation procedure (cf. section 3.4). The latter usually requires low temperatures to avoid denaturation. Otherwise, one would not normally expose protein solutions to organic solvents or to other denaturants such as urea or dodecyl sulfate.

## Catalytic Site Inactivation

The inactivation of enzymes due to a specific effect on the catalytic site is more difficult to avoid. Loss of cofactors can be prevented (if they are known) by adding them back or including them in the buffers used. A preincubation procedure with cofactor before assaying the enzyme may be needed; the apoenzyme is usually not less stable than the holoenzyme, so most activity lost can be restored. A more serious problem is covalent modifications of the active site due to exposure to the more hostile environment. The active site of an enzyme contains amino acid residues which are abnormally reactive, and so responsible for the enzyme's catalytic abilities. These residues are also more susceptible to modification. The most troublesome is cysteine; sulfhydryl residues at the active site may be in the ionized form, which is very prone to oxidation. Normally inside the cell the reducing atmosphere and the presence of other sulfhydryl-containing molecules protect these groups. Exposed to a higher oxygen tension, several fates of a sulphydryl are possible. These include:

(i) Disulfide bond formations:     $-S-S-$
(ii) Partial oxidation to a sulfinic acid:     $-S-OH$
(iii) Irreversible oxidation to a sulfonic acid:     $-S{\overset{\displaystyle\nearrow O}{\underset{\displaystyle\searrow OH}{}}}$

Formation of a disulfide bond requires another sulfhydryl to be in the vicinity and an oxidation process. Disulfide formation is greatly accelerated by the presence of divalent ions which can activate the oxygen molecule and complex with the sulfhydryls. Two protective actions can be taken: (1) removing all traces of (heavy) metal ions using a complexing agent such as EDTA; and (2) including a sulfhydryl-containing reagent in the solution, such as $\beta$-mercaptoethanol, or dithiothreitol/dithioerythritol (Cleland's reagents) (156):

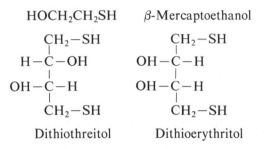

$$HOCH_2CH_2SH \qquad \beta\text{-Mercaptoethanol}$$

Dithiothreitol            Dithioerythritol

$\beta$-Mercaptoethanol is available as a heavy, odorous liquid. As a protective agent it should be used at a concentration of between 5 and 20 mM; lower concentrations quickly convert to the disulfide due to oxidation. The protective ability is lost within 24 hr unless the solution is kept in anaerobic conditions, as any disulfide that has formed can exchange with a still active sulfhydryl in the protein. Oxidized $\beta$-Mercaptoethanol can even accelerate the inactivation process (Figure 6.3).

Although disulfide bond formation is reversible, it is not a desirable thing to happen. The dithioanalogues of the reduced sugars threitol and erythritol are white solids, which if pure have little odor. A concentration of only 0.5 to 1 mM is sufficient to provide protection for 24 hr because the disulfide formed by oxidation is a stable intramolecular disulfide which does not interchange with protein sulfhydryls (Figure 6.4).

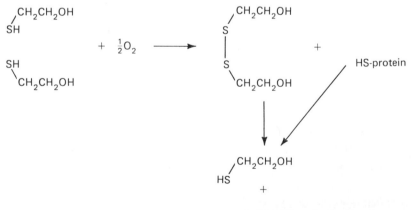

**Figure 6.3.** Inactivation of an active sulfhydryl in an enzyme by oxidized $\beta$-mercaptoethanol.

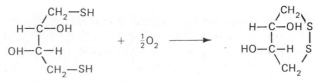

**Figure 6.4.** Oxidation of dithiothreitol to form the inactive cyclic disulfide.

Other sulfhydryl protective agents, of similar properties to mercaptoethanol, that have been used include 2,3-dimercaptopropanol, thioglycolate, glutathione, and cysteine. Cysteine oxidizes quite rapidly, especially in the presence of heavy metal ions.

A routinely successful procedure is to use mercaptoethanol diluted 1 part in 1000 ($\sim 12$ mM) during preparations (if required at all), and for longer-term storage dithiothreitol at 1–5 mM. EDTA is usually present in all buffers at a concentration of 0.1–0.2 mM. However, there are cases where the presence of a complexing agent is detrimental, since it may remove essential metal ions from enzyme active sites. Then very pure salts uncontaminated by heavy metals may be needed, or perhaps a weaker complexing agent (e.g., imidazole buffer) might suffice.

It is appropriate here to mention a few words about dilute enzyme solutions. It is well known that very dilute enzyme solutions lose activity quickly, but this loss can be prevented by the inclusion of a relatively high level of another protein, usually bovine serum albumin (BSA). In purification procedures one would not normally add a contaminating protein, so very dilute solutions should be concentrated as quickly as possible (cf. section 1.4). But for enzyme activity measurement it is likely that the final enzyme concentration during the assay may be as low as 1 $\mu$g ml$^{-1}$, which can lead to rapid inactivation; even diluting the sample just before assay may cause losses. In these cases the presence of BSA as stabilizer can be justified and should *always* be used. The mode of action of BSA is controversial, but a variety of effects may be involved. Extreme dilution of a protein may lead to dissociation of subunits which either may be inactive or, even if fully active, may be unstable and denature rapidly. It is possible that the BSA interacts directly with the dilute protein, either lessening the chances of dissociation or stabilizing the subunits. Also, a small amount of protein is likely to adsorb to the walls of the container, especially if it is glass. One $\mu$g of protein may all be lost on 5 cm$^2$ of container wall, but in the presence of 1 mg BSA, the 1 $\mu$g that adsorbed would nearly all be BSA, so little enzyme would be lost from the solution. For diluting protein solutions before assay, a suitable buffer containing as much as 10 mg ml$^{-1}$ BSA should be used, and the assay mixture itself should have at least 0.1 mg ml$^{-1}$ BSA.

## Proteolytic Degradation of Enzymes

Living cells contain the elements for their own destruction, namely digestive hydrolytic enzymes such as nucleases, polysaccharide hydrolases, phosphatases, and proteases (proteinases, proteolytic enzymes). In mammalian cells they are packaged in lysosomes and released under controlled conditions, usually targeting on specific molecules to cause their destruction as required by the physiological circumstances. In plants, digestive enzymes are stored in vacuolar spaces; their action in the cytoplasm may be under the control of a variety of specific inhibitors. In microorganisms, hydrolytic enzymes are often to be found between the plasma membrane and the cell wall, largely for the purpose of digesting extracellular macromolecules. The process for obtaining an enzyme extract almost inevitably destroys this delicate balance and allows these enzymes to mix with the cell contents. The proteolytic enzymes are of particular concern here. There are procedures for minimizing the disruption of lysosomes when extracting soft animal tissues (157), but in many cases these are not applicable when disruption of the cell itself requires vigorous homogenization. Also, on a large scale, hormonal or enzymic treatments to release cell contents gently are often not practical. Consequently we must accept the fact that digestive enzymes will be present and do something about them.

Proteolytic inhibitors are compounds, either chemical or biological, which specifically inhibit the proteolytic enzymes that occur widely in tissues. There are, unfortunately, many different classes of proteolytic enzymes, not all of which can be successfully inhibited. The main categories are (158):

(1) Serine proteases (e.g., trypsin, chymotrypsin, elastase-type activates), EC subclass 3.4.21
(2) Thiol proteases (e.g., papain, yeast proteinase B), EC subclass 3.4.22
(3) Acid proteases (e.g., pepsin, cathepsins D.), EC subclass 3.4.23
(4) Metalloproteinases (e.g., collagenases, many microbial neutral proteinases), EC subclass 3.4.24
(5) Carboxypeptidases (remove C-terminal residues), EC subclasses 3.4.16, 17
(6) Peptidases (mostly remove N-terminal residues), EC subclasses 3.4.11–15

The principle attacks on the proteolytic enzyme problem have been directed at the serine proteases, at first using the highly dangerous DFP (diisopropyl fluorophosphate)—dangerous because it is volatile and attacks another serine hydrolase of vital importance in nerve conduction, human acetyl cholinesterase. The introduction of phenylmethylsulfonyl fluoride (PMSF) (159), which reportedly does not inhibit acetyl cholinesterase, has made the enzyme chemists' attack on proteolysis a less dangerous occupation, and PMSF is now widely used. Although not very soluble, PMSF can be dissolved in acetone before dispersing into the extract to a final concentration of 0.5–1 mM. As it hydrolyses quite rapidly, it should be added directly to the solution containing the serine esterases, not dissolved in the buffer prior to making a homogenate. PMSF also inhibits some thiol proteases and some carboxypeptidases, so is a very useful reagent; once inhibited, the enzymes are dead, and no more PMSF is needed.

Acid proteases do not act via a serine or cysteine residue, and are unaffected by PMSF. However, fungi of *Streptomyces* spp. (160) produce an extracellular peptide of complex structure which is a potent inhibitor of acid proteases such as pepsin, cathepsin D, and yeast protease A. It is called "pepstatin A," and is effective at concentrations as low as $10^{-7}$ M. Pepstatin A is available commercially. A variety of similar proteolytic inhibitors have been described, including leupeptin, an inhibitor of some of the more active peptidases. These inhibitors bind tightly to the proteases, but they are reversible, so may need to be replenished during a purification procedure. Pepstatin A, attached to an agarose column (cf. section 4.5) has been used to purify cathpepsin D (161). One could consider passing an extract through a column containing a range of bound proteolytic inhibitors to remove the proteases before commencing the fractionation procedure. The class of metalloproteinases, which require divalent metal for activity, can be inhibited simply by complexing metal ions with EDTA.

Provided that the enzyme being isolated is not affected by any of the reagents, a suitable treatment for a freshly prepared extract is to include 2–5 mM EDTA, 0.5–1 mM PMSF and $10^{-7}$ M pepstatin A.

## Other Stabilizing Influences on Enzymes

Aqueous 1–2% solutions of proteins are not in a comparable state to the natural environment of the protein. Usually the nonaqueous components in the cell cytoplasm make up 10–15% w/v, and most of these components (mainly proteins) contain a substantial amount of bound or at least loosely associated water. NMR studies on the water of cells have indicated that much of the water may not be freely mobile. It is to be expected that proteins have evolved to be stable in these conditions rather than in the artificial solutions of a biochemist's laboratory. Consequently stabilization (against denaturation and other more subtle conformational changes that lead to inactivation) can be accomplished by attempting to mimic the conditions of relatively low water activity. The most widely used method is to include glycerol in buffer solutions. Glycerol, completely water-miscible, forms strong hydrogen bonds with the water, effectively slowing down the motion of the water molecules and so reducing the water activity. A recent analysis of the effect of glycerol suggests that the protein molecules preferentially bind water, and the structure so formed is less able to unfold against the structured glycerol solvent than it would be able to in water alone (162). Glycerol up to 50% w/v is used, though the highest concentrations are usually reserved for storage, since the viscosity of such a solution is too high to allow any manipulations other than electrophoresis. But 20–30% w/v glycerol mixtures do not have a great viscosity, and most procedures can be carried out in buffers containing this amount of glycerol. Ion exchange chromatography seems to be little affected by the presence of glycerol, which is a further indication that ion exchangers rely entirely on electrostatic interactions; hydrogen-bonding and hydrophobic forces, which would be weakened by glycerol, are not involved. Salting out may well be affected in terms of percentage saturation required for precipitation, and the precipitates may be impossible to centrifuge if the glycerol concentration is too high—the solvent is both too viscous and too dense.

Stabilization of enzymes with glycerol can be a dramatic effect; enzymes that otherwise die very rapidly can remain active for weeks, and because the solutions either do not freeze at all or form a thin slurry of ice and concentrated glycerol-solute mixture, storage at very low temperature is rarely detrimental. As an alternative to glycerol, sugar or sugar alcohol solution (glucose, sucrose, fructose, sorbitol) have been used successfully (163); the mechanism is presumably similar. Other hydrophilic solutes, e.g., formamide and dioxan, tend to be destructive to enzymes at high concentrations.

The removal of effective water activity seems to be generally beneficial provided that the means of doing it is not a destabilizing influence. High ammonium sulfate concentrations, where the enzyme might be precipitated, are very stabilizing, and in fact most enzymes sold commercially are as suspensions in 2–3 M ammonium sulfate. Alternatively, 50% glycerol or lyophilization is used to preserve activity—and as a bacteriostatic precaution. During enzyme purifications it is usually necessary to store fractions overnight or even for longer periods, and considerations must be given to optimize the storage conditions. If an ammonium sulfate fractionation is being carried out, leave the enzyme in as high a concentration of ammonium sulfate as possible, e.g., in a centrifuged pellet rather than after redissolving it. If storage in the absence of protective agents is unavoidable, consideration should be made of the relative advantages of short-term freezing or storage close to $0°C$ in the presence of proteolytic inhibitors. Freezing can inactivate enzymes. The susceptibility of different enzymes varies enormously. Losses will always be less if the initial protein concentration is high. It is never advisable to freeze a solution with protein concentration less than about 2 mg ml$^{-1}$. Remember that during freezing, first pure ice separates, which concentrates the protein and other solutes. Then the least soluble solute will precipitate out; if this is one component of a buffer, the pH will change markedly (cf. section 2.1). On freezing sodium/potassium/chloride/phosphate solutions the pH can change from neutral to 4 or lower; sodium salts aggravate the effect (4,5). The higher the protein concentration relative to buffer salts, the more it will be capable of acting as a buffer itself and counteracting radical pH shifts during freezing. Freezing at below the eutectic points of most salts ($-25°C$), is best, since even if there is a pH shift, true solidification will minimize any denaturation. Freezing at $-10$ to $-15°C$ is probably no better than not freezing at all, at least for short periods up to 24 hr.

# Chapter 7

# Optimization of Procedures and Following a Recipe

This chapter is intended for a later stage in purification efforts, when a method has been established or a published method has been tried but there may be some dissatisfaction with the overall procedure. The objectives of a scheme for the purification of an enzyme or other protein may be any or all of the following:

1. High overall recovery of activity
2. High degree of purity of the end product
3. Reproducibility
   (i) within the laboratory
   (ii) if scaled up or scaled down
   (iii) in other laboratories
4. Economical use of reagents and simple apparatus
5. Convenience in terms of working hours.

Although points 1 and 2 are obviously desirable, neither may be necessary for a particular application. If it is a novel procedure resulting in a useful preparation, the method may be publishable provided that point 3 is covered. For a publishable procedure point 3(iii) should always be attained—unfortunately, it is the most difficult to confirm! Also unfortunately, it is very often not attained, and many published procedures do not work very successfully in other laboratories for several reasons. Sometimes the originators did not repeat their methods often, and so their system was not as tightly defined as they thought. Occasionally there may be vital information missing from the published method. It is also the case that biological raw materials can never be identical; differences in extractability or composition can result in complete failure.

Points 4 and 5 above are technical considerations not related to the efficacy of the scheme itself, but to everyday organization and budgeting. The best procedure may be excellent in a well-equipped, well-funded laboratory, but out of the question otherwise. Also, whereas research students should be prepared to spend 12 or 16 hours non-stop if necessary, it is neither easy nor reasonable to persuade technical staff to do the same; if the process can be (literally) frozen midway and picked up the following morning, everyone would be happier.

Many of the comments below concern methods of cutting corners, of saving time or reagents in order to fit the exact requirements of the enzyme purification process. Mostly this means less than ideal operation, and is not for the purist. But speed can be beneficial, lessening the problems discussed in section 6.2, so even if less-than-ideal fractionation procedures are followed for the sake of speed, the end product could be of a better quality.

## 7.1  Speed versus Resolution: The Time Factor

Any process involving interaction of macromolecules is slow compared with the kinetics of small molecules—it may be in the order of minutes rather than seconds or milliseconds. Consequently at each stage of protein fractionation there is a need for equilibration; perhaps 20–30 min is needed to attain (almost) complete equilibrium. Suppose that a precipitation process has half-time of 2 min; the percent completion is plotted in Figure 7.1. Note that at 15–20 min the result is not measurably different from 100% completion, but at less than 10 min the process is insufficient to be considered "complete." The optimum time in this example would be around 15 min—longer is unnecessary, unless you need a cup of coffee. Generally one does not bother to find what the

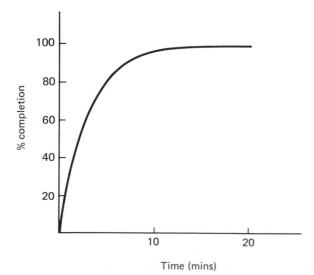

**Figure 7.1.** Percent completion of a first-order process with a half-time of 2 min.

optimum time is, but variation from one preparation to another could be due to cutting things too short.

Now consider what determines the time factors. First, most processes occurring are diffusion-controlled, whether aggregation of protein molecules, adsorption to a solid matrix, or movement in free solution. Halving the diffusion coefficient will double the time needed, and since the diffusion coefficient for any particular molecule is dependent on both temperature and viscosity of the medium, these two factors are very important. Traditionally enzymes have been considered so labile that they must always be kept at close to 0°C during isolation; most preparations are still done at low temperatures. Yet there are many—probably most—enzymes that are perfectly resistant to denaturation (except for organic solvent fractionation) at room temperature, especially in the *shorter time needed* to complete the fractionation. The 20°C difference between cold-room temperature and average laboratory temperature typically increases chemical reaction rates by three- to four-fold. At 25°C any oxidative, proteolytic, or other detrimental phenomenon other than denaturation can be expected to occur that much faster than at 4–5°C, but by the same token any fractionation procedure can be expected to take *less* time to reach equilibrium at 25°C. This is of particular importance for column techniques, since the maximum speed of operation of ion exchange or gel filtration columns is determined by approach to equilibrium during the chromatography. If the diffusion coefficient for a protein in the solvent is 4 times lower at 5°C than at 25°C, then the flow speed of the column must be 4 times less. However, the viscosity of water is 1.8 times higher at 5°C than at 25°C; this is the main contribution to diffusion coefficients being smaller at lower temperatures; a factor of 2 is more probable than 4 for diffusion. Consequently a halving of speed, but a quartering of the rate of inactivating phenomena, makes cold operation more desirable. Often people working with enzymes will happily leave a solution in the cold overnight, yet would consider leaving it at room temperature for an hour or so as fatal. Now 4–5 hr at room temperature is probably equivalent to overnight in the cold; unless one is sure of the stability of the preparation, neither should be allowed to occur except as a necessary part of the fractionation procedure.

Stabilization of enzymes with glycerol or other polyhydroxylated compounds has been found necessary in many cases. Levels of glycerol from 20 up to 50% w/v have been used. Pure glycerol is nearly 100 times more viscous than water, though the relationship is fortunately not linear for mixtures. Some values for the viscosity of glycerol and sucrose solutions are given in Table 7.1. Note that 25% glycerol is twice as viscous as water, 50% glycerol 6 times as viscous. Thus processes necessitating glycerol as a stabilizer require extra time in these proportions. A column run in 25% glycerol at 5°C should be run 4 times slower than in the absence of glycerol at 25°C. Obviously the stabilization factor obtained by using glycerol must be higher than this time factor to make it worthwhile.

What are the times needed for almost complete equilibration? No firm figures can be given as each case obviously is different (large proteins diffuse more

**Table 7.1.** Viscosity of Sucrose and Glycerol
Solutions

| | Viscosity (cP) | | |
|---|---|---|---|
| | Sucrose | | Glycerol, |
| % w/v | 0°C | 25°C | 25°C |
| 0 | 1.78 | 0.89 | 0.89 |
| 10 | 2.46 | 1.18 | 1.06 |
| 20 | 3.77 | 1.70 | 1.54 |
| 30 | 6.66 | 2.74 | 2.12 |
| 40 | 14.58 | 5.16 | 3.18 |
| 50 | | | 5.2 |

slowly than small ones, which might involve another factor of 2). But in general
the following rule-of-thumb figures can be used (Table 7.2).

The total time taken for a fractionation step is theoretically independent of
volume, but in practice large volumes inevitably take longer to deal with. This
is discussed further in section 7.3. Remembering that the faster an isolation
procedure is carried out, the less likely the final product is to be degraded, some
decisions about compromising on the ideal conditions may need to be taken.
For instance, the following might apply: "If I run the column in the cold room
at optimum speed, the enzyme should be off at about 8:00 P.M., and the enzyme
can be left in the fraction collector tubes until morning. On the other hand, if
I run it at room temperature and faster than optimum speed, resolution may
be somewhat worse, but it will be finished by 4:00 P.M., giving me time to concentrate the preparation and freeze it away." In this example, the cold-room

**Table 7.2.** Approximate Times and Flow Rates Needed to Achieve Close to
Equilibrium Conditions in Protein Fractionations

| Method | 5°C | 25°C |
|---|---|---|
| Ammonium sulfate fractions | 20–30 min | 10–15 min |
| Organic solvent precipitation | 15–20 min | (Not relevant) |
| Isoelectric precipitation | 20–30 min | (10–15 min)[a] |
| Ion exchangers: CM-, DEAE-celluloses | 10–15 cm hr$^{-1}$ | 20–30 cm hr$^{-1}$ |
| CM-, DEAE-agaroses; Sephacel | 15–20 cm hr$^{-1}$ | 30–40 cm hr$^{-1}$ |
| Affinity adsorbents[b] | 5–10 cm hr$^{-1}$ | 10–20 cm hr$^{-1}$ |
| Immunoadsorbents | Contact times of several hours | |
| Dye ligand adsorbents[b] | 10–15 cm hr$^{-1}$ | 20–30 cm hr$^{-1}$ |
| Gel filtration—fine-grade dextran beads, agarose, Ultrogel | 2–5 cm hr$^{-1}$ | 4–10 cm hr$^{-1}$ |
| Sephacryls | 10–15 cm hr$^{-1}$ | 20–30 cm hr$^{-1}$ |

[a]Usually done cold to avoid denaturation at unphysiological pH.
[b]"Kinetic adsorption" has been run at up to 500 cm hr$^{-1}$ (12) for very tightly binding proteins.

operation would involve 24 hr at 4°C before progressing to the next step. At room temperature only 5–6 hr may pass before storing away the sample in a safe manner. On the one hand, the cold preparation may be more degraded by proteolysis and inactivated by prolonged existence in dilute solution; on the other hand, the room-temperature preparation may be rather more contaminated with unwanted proteins. The object at this stage is not to recommend one procedure over the other; it is just to point out the alternatives.

Some enzymes are remarkably resistant to any inactivation by all the phenomena described in section 6.2. In many cases this is just as well, as purification procedures taking weeks to complete have been reported. Extracellular enzymes are more resistant because their natural environment is harsher. In these cases speed should be of little concern once proteolytic problems have been solved. But in other cases where stability is limited, speed of operation can be much more important than resolution at each step, even if it means putting in an extra step to remove final impurities. Processes to be avoided are dialysis and gel filtration on slow media. If it is possible to arrange, one fractionation step should be able to follow from another without extensive preparation. Thus salt fractionation may be followed by a gel filtration step, since the gel filtration will remove the salt as well as fractionate the proteins. But it cannot be followed directly by an ion exchange technique except in exceptional cases where the residual salt does not prevent adsorption. An ion exchange step cannot be followed directly by gel filtration, because the fraction would need concentrating—it is rare that ion exchange elution is obtained in sharp, concentrated peaks (but cf. discussions of affinity elution, section 4.4, and chromatofocusing, section 4.3). Organic solvent fractionation may be followed by ion exchange, although if the enzyme is only weakly adsorbed, endogenous salts co-precipitated by the organic solvent might increase the ionic strength too much to allow adsorption. A summary of these direct sequential fractionations is given in Table 7.3, together with the processing that can be done between fractionations to allow any sequence of steps. It is worth noting that in many cases, especially preceding a chromatographic step, a clarifying centrifugation is needed before proceeding.

Some possible chains of steps which do not involve intermediate treatments are:

Salt → Gel filtration → Ion exchange

Organic solvent → Affinity adsorption, dye ligand, etc.

Organic solvent → ⎡ Ion exchange ⎤ → Salt → Gel filtration
⎣ Affinity adsorption ⎦

Affinity adsorption → Salt → Gel filtration

Remember the capacities of each technique; precipitation methods are particularly suited to the first step, as very large amounts of material can be dealt with. Gel filtration has a moderate capacity unless one has giant columns.

**Table 7.3** Possible Sequences of Fractionation Processes, Indicating Which Require Intermediate Processing

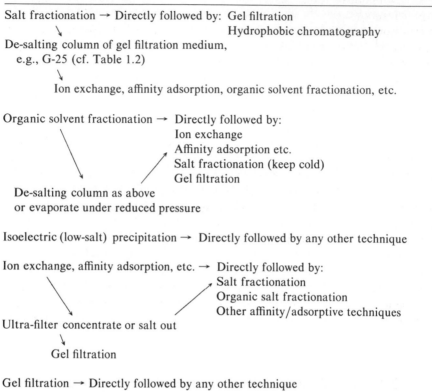

Salt fractionation → Directly followed by:  Gel filtration
         ↘                                  Hydrophobic chromatography
De-salting column of gel filtration medium,
    e.g., G-25 (cf. Table 1.2)

            ↘
        Ion exchange, affinity adsorption, organic solvent fractionation, etc.

Organic solvent fractionation →  Directly followed by:
                                  Ion exchange
                                  Affinity adsorption etc.
                                  Salt fractionation (keep cold)
                                  Gel filtration
    De-salting column as above
    or evaporate under reduced pressure

Isoelectric (low-salt) precipitation →  Directly followed by any other technique

Ion exchange, affinity adsorption, etc. →  Directly followed by:
                                            Salt fractionation
                                            Organic salt fractionation
                                            Other affinity/adsorptive techniques
Ultra-filter concentrate or salt out

        Gel filtration

Gel filtration →  Directly followed by any other technique

Affinity adsorption has a low capacity, although one may be able to process a lot of protein if very little actually adsorbs. "De-salting" can be quite rapid, and the following schemes are convenient for routine use:

Salt fractionation → "De-salting" →
Ion exchange or affinity adsorption →
"Salting-out" concentration → Gel filtration

or

Organic solvent → Salt fractionation →
"De-salting" → Ion exchange or affinity adsorption

If these three or four steps are sufficient, it is often possible to complete the procedure in 5–6 hr, and if the crude extract is available early in the day, a 1-day preparation is achieved. But pauses can be tolerated, provided that degradative changes are prevented as far as possible. Two main storage processes for overnight or even longer are recommended. These are as follows: (i) Storage of an ammonium sulfate pellet in the cold (not frozen)—preferably before dissolving it with buffer. The high salt concentration stabilizes the enzyme and generally inhibits proteases. However, this is not an indefinite storage for impure fractions, and 2–3 days at 4°C is the longest that it should be kept. (ii) Frozen storage. If the temperature is low enough, all degradative processes will stop and the sample can be kept indefinitely. A temperature below −50°C is recommended. Normal deep-freeze temperatures are usually suitable for overnight storage, but not above −15°C. However, do not try to freeze a dilute protein solution; dilute solutions ($<2$ mg ml$^{-1}$) are much more likely to denature on freezing and thawing (dilute solutions also take up more room in the freezer!). Take note of the pH before freezing and make sure the solution is well buffered. Freeze in plastic containers (glass containers break when ice formation increases the volume), and thaw rapidly in warm water. Careful use of a microwave oven may be justified on a large scale. Remember that plastic is a poor heat conductor, and even if the water temperature is 50°C, the inside surface will not be higher than 30°C, and with agitation there is little chance of heat denaturation during thawing.

In summary, the time factor can be very important when dealing with labile enzymes or with extracts containing much proteolytic activity. It may well be better in the long run to rush one step after another and sacrifice some resolution, rather than carry out each process to perfection but take a long time over the isolation. On the other hand, with stable enzymes, attention to maximum resolution may be best, especially if it eliminates the need for further steps to remove impurities.

## 7.2   Scaling Up and Scaling Down

Trial procedures are carried out on a small scale, with the intention of increasing the scale when a method has been worked out. Often enzyme purifiers have been frustrated because something went wrong scaling up their method and things did not behave as expected. This is particularly annoying when much effort is needed to obtain the raw material, and the whole fractionation may have to be abandoned. It is useful to discuss the sorts of things that can vary when scaling up and what one should do, at least theoretically, to minimize these variations.

The first source of variability is the extract. It is likely that in order to process more raw material, a different, larger homogenizer will be used. The time needed to disrupt all the material may be different, and so the composition of the extract may differ, too. Even if the amount of enzyme extracted is comparable with the small-scale method, the amounts of other proteins and nonprotein material may be different, and this can have an effect on the precise fractionation conditions needed. Moreover, different levels of contaminating proteins passing through to the second, third, and later steps in fractionation may upset things. If there is a problem here, a small-scale trial on the large-scale extract is advisable.

Scaling up of precipitation steps should not be a problem. But it is worth noting that a precipitate after centrifugation contains a proportion of the supernatant fluid entrained in it, the proportion depending on the tightness of packing of the precipitate. Small-scale trials will often use centrifugation at quite high speed (since high $g$ forces are available with small rotors), whereas scaled up this might not be possible; the precipitate could be less tightly packed. And so a larger proportion of contaminants would be carried over into the next step. Even more important is the amount of precipitant carried over; for instance, an excessive amount of ammonium sulfate could upset a gel filtration column; proteins might even re-precipitate out on the column. If a soft precipitate is a problem, transfer of the precipitate slurry, after decanting most of the supernatant, to a small centrifuge followed by high-speed centrifugation can be the solution.

Scaling up of column chromatography needs very careful consideration. Important factors are the amount of protein applied per unit volume of adsorbent and the flow rate. Theoretically, a scaling up of 10-fold should be done by increasing the cross-sectional area 10 times, keeping the height the same, and also increasing the flow rate 10-fold. However, as discussed previously, short, squat columns present problems of uneven flow, and it is better to compromise by increasing *both* cross-section *and* height. Volume should be increased proportionately with the amount of protein applied; the best compromise is to keep the two-dimensional shape viewed from the side in the same proportions. Thus for a 10-fold volume increase, a column of 5 cm height by 1 cm diameter could be increased to 12 cm height by 2 cm diameter (assuming that columns are not available in all possible diameters). The flow rate, expressed in cm $hr^{-1}$, should be the same as in the small-scale trial. Since the cross-sectional area is 4 times greater, the flow rate should be only 4 times faster *expressed in ml $hr^{-1}$*. Thus the running time will be 2½ times as long. With ideal buffer flow, resolution and behavior should depend only on the relationship between column volume and applied sample. At a fixed flow rate (cm $hr^{-1}$) the column shape should not matter.

Major differences occur in scaling up when temperature alteration is needed. It is easy to heat 100 ml to 50°C in a couple of minutes, hold it there for 5 min, then cool rapidly. But for 2000 ml the heat transfer problems inevitably mean slower attainment of the temperature and slower cooling. So, if there is a heat denaturation step, make sure that in those critical minutes just before and after the maximum temperature there is not more denaturation than intended. Whereas rapid heating can be achieved with a microwave oven, the problem of rapid cooling remains. Do a trial on a small scale with slow heating and cooling. Similarly, if a procedure calls for cooling to 0°C before adding organic solvent, remember that it can take a very long time to chill several liters of solution. Use glass containers, since plastic is such a poor heat conductor. On the other hand, once the low temperature has been reached, a larger volume stays there easily.

Scaling up is really just a matter of common sense; it is bound to increase the time of operation, so plan carefully. Similarly, scaling down allows for more rapid operation, and this is worth considering. With centrifugations lasting only 5 min, and columns running within 1 hr, small-scale procedures can be completed in a very short time. So if not much final product is needed, scaling down is worth considering. The following describes a purification of fructose 1,6-bisphosphate aldolase from a rabbit muscle extract (the extract can be stored for long periods at $-30°C$).

Extract thawed: 50 ml taken at time zero; pH adjusted to 6.0; operations carried out at 25°C:

Dissolve in 12 g ammonium sulfate (time taken, 2 min); stir 5 min; centrifuge 20,000 $\times$ $g$ for 3 min.

Dissolve into supernatant 6 g ammonium sulfate; stir; centrifuge as above; dissolve precipitate in 5 ml buffer.

De-salt into K-Mops pH 7.0 buffer on 50-cm$^3$ G-25 column (10 min).

Run on CM-cellulose 4 cm$^2$ $\times$ 4 cm, wash with K-Tes buffer pH 7.5 (10 min).

Add 0.2 mM fructose 1,6-bisphosphate to washing buffer: aldolase is eluted; add 0.5 g ammonium sulfate per ml of enzyme (3 min); stir 5 min; centrifuge; resuspend precipitate for crystallization.

Time elapsed (min)

| | |
|---|---|
| 0 | Ammonium sulfate dissolved in;<br>stir;<br>centrifuge. |
| 20 | Ammonium sulfate dissolved in;<br>stir;<br>centrifuge. |
| 40 | Dissolve precipitate;<br>de-salt. |
| 60 | Run on CM-cellulose;<br>wash with pH 7.5 buffer. |
| 80 | Elute with fructose 1,6-P$_2$;<br>salt out with ammonium sulfate. |
| 100 | Centrifuge;<br>resuspend precipitate for crystals. |

The four processes, salt fractionation, de-salting, ion exchange with affinity elution, and salting-out concentration, are condensed into 2 hours of work. This is a particularly easy example, but many other purifications can be carried out very rapidly on a small scale. It is hardly worth keeping things cool, especially as this would delay things somewhat (cf. section 7.1).

It is sometimes necessary to scale down from a published procedure, when the originators presented their preparation on a vast scale. For instance "110 kg horse muscle was . . ." (!) (164). In these cases one must take into account the same considerations as discussed for scaling up, in reverse. Mostly, when trying to follow a published procedure, small-scale trials will be carried out, so there is a double problem of scaling down and assuming that the published procedure will work. This is discussed in the next section.

## 7.3   Following a Published Procedure

This section will apply to most people; there are so many published procedures, often a dozen or more for one enzyme, that it is unwise to set off without consulting previous work. In fact, many people will attempt to follow a published method exactly. Often it works well, and there is no need to make any but minor adjustments to suit one's own situation. But equally often a published procedure proves to be quite unreproducible, perhaps failing utterly to result in any usable enzyme at the end. Sometimes this is the fault of the authors, sometimes it is due to carelessness. The purpose of this section is to describe what approaches should be taken both at the outset and later when something goes wrong.

First of all it is important to distinguish trivial information from the important facts. The examples below are taken from actual published procedures:

> The pellet was extracted with 469 ml buffer A . . . and dialysed overnight against buffer A to remove ethanol.

The precise number of ml to three significant digits is not important; 450 or 500 ml would give equally good results—the volume will change on dialysis anyway.

> . . . eluted with a linear gradient of NaCl; fractions of *about* 3 ml were collected. . . . the active fractions (41 to 53) were pooled.

The tube numbers are not relevant to the reader since the authors have not defined exactly what volume had passed through the column before the activity emerged.

> . . . dialysed against buffer B for 24 hr.

Was there any change of dialysate—what volume of buffer B was used? These could be vital to the completion of dialysis and the behavior in the next step.

> The mixture was heated with continuous gentle stirring for 10 min at 62°C.

How long did it take to reach 62°C, and how quickly did it cool?

> The debris was spun down in an MSE Mistral 62 centrifuge at 2000 rpm for 20 min at 5°C.

You do not need to go out and buy an MSE Mistral 62 centrifuge in order to reproduce this step; the authors should have indicated the relative $g$ force that 2000 rpm in this machine represents.

> ... CM-cellulose equilibrated with 5 mM Tris-HCl pH 6.5.

This is extraordinary, as Tris has no significant buffering power at 6.5, especially at only 5 mM. This step needs watching carefully, and may need modification.

> ... with 50 mM Na acetate buffer, pH 5.0.

How was this buffer made? Was it 50 mM Na acetate adjusted with acetic acid, or with HCl, to pH 5.0? Definitions of buffer composition are not always clear. But then with buffers, how important is it going to be? For ion exchange chromatography, ionic strength is equally important as the pH, and the buffer composition must be exactly defined. Otherwise, provided that the pH is correct, the buffer strength and composition may not matter much. Even the nature of the buffering species should not matter; provided it has a similar $pK_a$, another buffer might do quite well, especially if the suggested one is rather exotic and not immediately available. Thus cacodylate has largely been superceded by $N$-morpholinoethane sulfonate (Mes). $N$-Ethylmorpholine ($pK_a$ 7.7) is rarely more advantageous than Tris ($pK_a$ 8.1) or triethanolamine ($pK_a$ 7.8). Diethyl barbiturate (called Veronal in older work), $pK_a$ 8.0, $n = -1$, is also not usually used now; Tricine is a better equivalent, if an anionic buffering species is needed. But remember that some buffers complex metal ions, and this might be important in stabilizing the enzyme (cf. section 6.1). It has already been mentioned that the pH of a buffer solution needs definition of the temperature at which the pH was measured, and although authors often say "all operations were carried out at $4°C$", it is a fair bet that in most cases the pH values of buffers were adjusted at room temperature.

> To each 100 ml of extract, 49 ml (0.25 saturation) of saturated ammonium sulfate was added.

The comment in parentheses is clearly mistaken: 49 ml of saturated added to 100 ml makes $49/149 = 0.33$ saturation. This same error appeared several times in the paper, and clearly the authors were confused as to what fractional saturation means—or at least they had their own secret definition. It is fortunate that the double information was given, since it is likely that the part not in parentheses is what was actually done, and can be repeated. If they had only said "saturated ammonium sulfate was added to 0.25 saturation," then their mistake would have resulted in irreproducibility. The literature abounds with peculiar definitions (e.g., "25% saturation" of ammonium sulfate has been used to mean 25 g/100 ml), and one must be careful to be sure what was really intended.

After successfully reproducing a published purification procedure, one should have some feel for any dubious steps and whether some improvements may be made. A decision to try something different may be very reasonable if the procedure was developed before some new reagent or method was available, which might make the whole process simpler. But it is unwise to make minor variations on the reported steps unless one is very unhappy about, for instance, the percentage recovery, since most probably the authors have already spent some time making variations, and published what they considered was the optimum procedure. If uncertain about some details, the quickest solution is to contact the author direct.

# Chapter 8
# Measurement of Enzyme Activity

## 8.1  Basic Features of Enzymic Catalysis

Most people approaching the problem of enzyme purification will be following a standard procedure for measuring enzyme activity. But in some cases, it may be useful to develop an alternative assay, for instance, a rapid spectrophotometric method to avoid queuing for the use of a scintillation counter. On rare occasions a newly discovered enzyme will need a new assay method. Ultimately the biochemist wants to know the relationship between an enzyme's activity in vitro and its physiological situation. But the pH, ionic strength, and, particularly, the substrate concentration in vivo are likely to be quite different from the conditions used to measure the enzyme activity during a purification process. What is required here is a reliable, reproducible assay which will relate the amount of enzyme in one fraction with another, regardless of what the "activity" may mean in a physiological context.

Enzyme activity is affected by the concentrations of its substrate(s), activators, and inhibitors specific for the enzyme, nonspecific effects of compounds such as salts and buffers, pH, ionic strength, temperature, and in some cases interactions with other proteins or membranous material that might be present. Usually one aims at having all conditions optimal, so that the enzyme exhibits maximum activity ($V_{max}$). But this is not always possible, for reasons of substrate cost or solubility, or perhaps compatibility with conditions required by coupling enzymes for activity detection. These conditions will be discussed below.

## Substrate Concentration, Activators and Inhibitors

Activity is nearly always greatest at the highest feasible substrate concentration. If the enzyme obeys Michaelis–Menten kinetics, then a substrate concentration of at least 10 times the $K_m$ should be used. At $10 \times K_m$, the rate is 91% of the theoretical rate at infinite concentration. Remember that the concentration of one substrate will usually affect the $K_m$ of another in a way that depends on the type of reaction mechanism. For instance, for many sequential mechanisms, the higher the concentration of substrate B, the smaller is the $K_m$ for A, and vice versa. So a high concentration of one of the substrates (e.g., the cheapest) would mean that less of the other is required to achieve a concentration of $10 \times K_m$. On the other hand, nonsequential (ping-pong) mechanisms have the peculiar property that the higher the concentration of one substrate, the higher the $K_m$ (lower the apparent affinity) for the other, so to approach $V_{max}$ both substrates may be needed at quite high concentration.

There are two reasons for operating well above the $K_m$ value. First, slight variation of concentration from one set of assays to another makes very little difference. A 10% error may result in only 1% change in rate, barely detectable. Thus the concentration of the substrate, which may be difficult to determine, need not be known accurately. Second, consumption of substrate during the assay will make little difference to the rate (Figure 8.1), and, if products are accumulating, any inhibitory effect they may have is less if the substrate concentration is high (Table 8.1). In the example shown in Table 8.1, the product binds to the enzyme as tightly as the substrate; if such a product is not removed during the assay, the rate will decline steadily and the true *initial* activity will be underestimated. Readily reversible enzymes such as isomerases fall into this category. It is sometimes technically convenient to measure an enzyme's activity in the direction where $\Delta G^0$ is positive, i.e., the equilibrium favors the substrates; in such cases removal of *all* products during the assay is

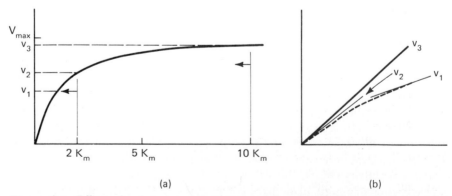

**Figure 8.1.** Effect of substrate consumption on enzyme activity: (a) relationship between rate and substrate concentration; (b) observed rates as substrate is consumed, starting with [substrate] = $10K_m$ (solid line) and [substrate] = $2K_m$ (dashed line).

**Table 8.1.** Rate of Enzyme under Different Situations of Substrate Concentrations

| | Concentration of substrate | Concentration of product | Rate (% $V_{max}$) |
|---|---|---|---|
| Initial concentration | $10K_m$ | 0 | 91 |
| Final, with product removed | $9K_m$ | 0 | 90 |
| Final, with product accumulating | $9K_m$ | $1K_{ip}$ | 82 |
| Initial concentration | $2K_m$ | 0 | 67 |
| Final, with product removed | $1K_m$ | 0 | 50 |
| Final, with product accumulating | $1K_m$ | $1K_{ip}$ | 33 |

*Key:* $K_m$ = Michaelis constant for substrate; $K_{ip}$ = inhibitor concentration for product. $K_{ip}$ is assumed to be equal to $K_m$.

desirable. Removal of products is a natural result of coupled assays (cf. section 8.3), but inclusion of systems to remove products even in a stopped assay (cf. section 8.2) can be useful.

Any assay takes a finite time to complete; disregarding the attainment of the initial steady state, which takes only milliseconds, the rate should be linear from zero time up to the time that the product formation can be observed or measured accurately. To overcome problems of product inhibition and approach to equilibrium, various strategies can be used. These include using very high concentrations of substrate, unphysiological pH (if a proton is involved in the reaction), or use of a very sensitive method for detecting product, so that little conversion is necessary during the course of the assay. As an example, the enzyme malate dehydrogenase operates in both directions physiologically:

$$\text{Malate}^{2-} + \text{NAD}^+ \rightleftharpoons \text{Oxaloacetate}^{2-} + \text{NADH} + \text{H}^+$$

The equilibrium is far to the left at pH 7; nevertheless, a high $\text{NAD}^+/\text{NADH}$ ratio in cells, combined with rapid removal of oxaloacetate, can result in the net reaction progressing from left to right. Inside the cell, of course, it will simultaneously be going in the other direction, it is just that overall *more* catalysis in the direction of oxaloacetate formation might be occurring than in the other direction. A lot of the enzyme present would be taken up with catalysing the reverse direction, so the next reaction (flux) would be far less than the $V_{max}$ condition and would bear little relationship to the amount of enzyme present. Since one wants to know the amount of enzyme present, conditions must be arranged such that reversal of reaction does not occur. Malate dehydrogenase activity can be detected directly by formation of NADH (or, in the reverse direction, loss of NADH), because NADH absorbs at 340 nm, whereas $\text{NAD}^+$ does not (Figure 8.2). The enzyme can be assayed in the thermodynamically favorable direction using oxaloacetate plus NADH at a pH of 7 or somewhat less, in which case a certain value of activity is obtained. Alternatively, it can be forced to go in the forward direction by (i) a high concentration of malate, typically 50–100 mM; (ii) a high pH to remove protons as they are

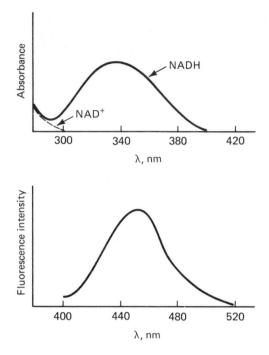

**Figure 8.2.** (a) The absorption spectrum of NADH, compared with that of NAD$^+$; (b) the fluorescence emission spectrum of NADH, with excitation at 365 nm.

formed—pH 9–10; and (iii) removal of oxaloacetate by trapping as a hydrazone:

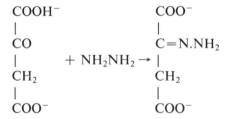

If all three of these conditions are used, then the enzyme's activity can be measured as NADH formation. But note that there is no reason why the activity in absolute units should be the same in this direction as in the other. With physiologically reversible enzymes they are often similar; in the case of malate dehydrogenase the maximum rate attainable in the forward direction is about half of that in the reverse direction.

The conventional picture of the effect of substrate concentration, hyperbolic kinetics with an asymptote approaching $V_{\max}$ (Figure 8.1a), may not apply if either allosteric effects occur or the phenomenon of "substrate inhibition" occurs. With allosteric enzymes, assay conditions are chosen so that any activators needed are present and substrate concentrations are high—often under these conditions allosteric behavior is abolished. *Substrate inhibition* is the term used to describe the effect of reduced activity at high substrate concentration. In the case of malate dehydrogenase (above) assayed in the direction of malate formation, oxaloacetate concentrations higher than about 0.2 mM lead to a lowering of the rate, due to the formation of a tight oxaloacetate-$NAD^+$-enzyme abortive complex. In other cases the reasons for substrate inhibition are not at all clear. Yeast pyruvate kinase has been assayed at very high concentration of substrates (phosphoenol pyruvate and ADP) and activator (fructose-1,6-bisphosphate), 10 mM of each—up to 100 times the $K_m$ values (165). However, at 1 mM of each, a higher activity is obtained—and at much less expense! Now 10 mM each of phosphoenol pyruvate, ADP, and fructose-1,6-bisphosphate have a combined theoretical ionic strength of 0.26 M, which together with the buffers present make a total ionic strength of 0.4 M, well above physiological; it is probable that this resulted in the apparent "substrate inhibition." When dealing with multiple-charged substrates, their effect on ionic strength should be considered.

It is also possible that a substrate preparation can include an inhibitor. Unfortunately, if it is a competitive inhibitor (most likely), this will not be detected except by comparison with another, better grade of substrate. Mathematically, one can describe these situations as follows:

For a competitive inhibitor:

$$\frac{V_{max}}{v} = 1 + \frac{K_m}{s}\left(1 + \frac{i}{K_i}\right)$$

where $v$ = rate observed; $s$ = substrate concentration; $i$ = inhibitor concentration; $K_i$ = enzyme-inhibitor dissociation constant.

If the inhibitor is present in the substrate, then $i \propto s$,

$$\frac{V_{max}}{v} = 1 + \frac{K_m}{s} + \beta$$

where $\beta$ is a constant.

At infinite $s$, $V_{max}/v = 1 + \beta$; thus both the determined $V_{max}$ and $K_m$ are less than the true values by a factor of $1/(1 + \beta)$, but hyperbolic kinetics apply (Figure 8.3).

If the inhibitor is mixed noncompetitive:

$$\frac{V_{max}}{v} = 1 + \frac{K_m}{s} + \frac{i}{K_i} + \frac{K_m}{s} \cdot \frac{i}{K_i'}$$

where $K_i'$ = dissociation constant of enzyme-substrate-inhibitor complex.

Putting $i$ proportional to $s$:

$$\frac{V_{max}}{v} = 1 + \frac{K_m}{s} + \alpha s + \beta$$

Thus a plot of $1/v$ against $1/s$ would be nonlinear, and apparent substrate inhibition would occur (Figure 8.3).

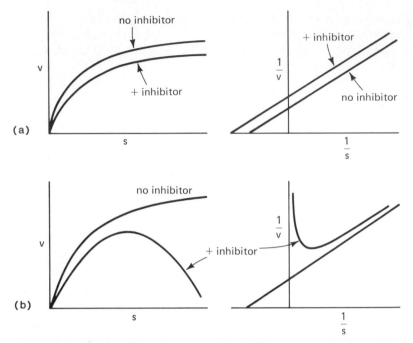

**Figure 8.3.** (a) The effect of a competitive inhibitor present in the substrate preparation; rate versus substrate, and reciprocal plots; (b) as (a), but with a noncompetitive inhibitor in the substrate.

An example of a competitive inhibitor going unrecognized for a long time is the recent discovery that aluminum contamination in commercial ATP solutions is responsible for variable activities of yeast hexokinase (166). The Al · ATP complex is a potent competitive inhibitor ($K_i \sim 0.1 \, \mu M$), but hyperbolic kinetics still apply; $V_{max}$ and $K_m$ for ATP are both reduced. The presence of a noncompetitive inhibitor can be detected as indicated above; the inhibitor may be salt, an undesirable metal ion, an organic solvent that the substrate is dissolved in, or perhaps a compound more subtle.

In conclusion, there are a number of pitfalls when choosing a suitable substrate concentration. Questions of economy may override ideality; in any case, *routine* assays are not necessarily required to be absolutely correct and accurate.

## pH, Ionic Strength, and Temperature

Having stated that substrate concentrations should be, if possible, at least 10 times the $K_m$ values, it should be made clear that $K_m$ values are dependent on other influences, such as pH, ionic strength, and temperature. Similarly, the catalytic rate constant $k_{cat}$ (where $V_{max} = k_{cat} \times$ total enzyme) is dependent on these variables. Thus, although the optimum pH is regarded as that pH value (or range of pH values) where a maximum activity is obtained (maximum $V_{max}$), lowering activity with changing pH could be due to a sharply rising value of $K_m$ (and so decreasing saturation) rather than a decline in $V_{max}$. Some enzymes have optimum pH values well away from the physiological value, when measured at high substrate concentration. For instance, alkaline phosphatases, as their name implies, may be optimally active at pH 10–11, yet naturally operate at a pH close to 7. This is because physiologically, $K_m$ is much more important than $V_{max}$; the natural substrates occur only at very low concentrations. The $K_m$ in this case is much lower at pH 7 than at pH 10.

This digression serves to point out that an "optimal pH" must be very carefully defined; the enzymologist refers to the optimal pH as that pH where $k_{cat}$ ($V_{max}$) is highest—i.e., independent of substrate concentration. To establish a reproducible assay for enzyme purification these subtleties are unimportant; using a suitably high substrate concentration, the rates at a range of pH values can be tested, and the optimal value of $v$ (not necessarily $V_{max}$) is found. The nature of the buffer can be very important here, since many buffers can upset the behavior by directly or indirectly acting as inhibitors. Phosphate or other multiple-charged anionic buffers such as citrate may compete with a negatively charged substrate by binding weakly to the positively charged active site. Since buffer concentrations used are typically between 10 and 100 mM, even a weak association may inhibit significantly. Metal ions essential for the enzyme's activity may be complexed by the buffer; particularly effective metal-complexing buffers are citrate and histidine. Finally, the ionic strength of the buffer plus any other salts present could affect the enzyme's activity. Physiological ionic strengths are normally in the range 0.1–0.2 M. Enzymes are not often less active at lower than the physiological value, but as salt concentration increases above 0.2 M, activity is often depressed. (Remember when testing an ammonium sulfate supernatant not to use much sample, unless it has been established that this salt is not inhibitory.)

Finally, one must consider the temperature at which assays are carried out. There is no such thing as an optimum temperature of an enzyme; there is only an optimum temperature for a given assay method. Enzyme reaction rates, like almost every other chemical process, increase with temperature, typically by a factor of between 1.5 and 2.2 for every $10°C$ (1.8 is a good mean value). At high temperatures the counteracting force of protein denaturation means that, in the finite time needed to determine the activity, some enzyme is lost by denaturation, so increasing temperature eventually leads to decreasing product formation. The shorter the incubation time, the higher will be the *apparent* optimum temperature, since there is less time for enzyme to denature. Typical apparent optimums for assays taking 5–10 min to complete are in the range $40$–$60°C$. Much higher values are obtained for enzymes isolated from thermophilic organisms.

For a routine assay it would be inconvenient and (except for the thermophilic enzymes aluded to above) unrealistic physiologically to use high temperatures. Moreover, some standardization of temperature would be a desirable principle. The standard temperature in chemistry has stood for a long time at 25°C (298 K), and at first recommendations were that enzymes should be assayed at this temperature. However, there were complaints from workers in tropical areas (and in pre-oil-crisis days, from the United States) that their laboratory temperature was usually higher than this, and cooling systems were not readily available. So the Enzyme Commission in the 1964 edition of its manual[1] recommended 30°C as the standard temperature for enzymes—which was in any case closer to the 37°C temperature that so many enzymes were naturally exposed to. Later recommendations have omitted references to temperature in defining enzyme activity, and there has been a recent tendency to return to 25°C as a standard. Sometimes it is not possible or convenient to maintain the temperature at exactly the value demanded, in which case a correction can be applied. The value of $Q_{10}$ (activity at $T + 10°C$ divided by the activity at $T°C$) is easily determined. Use a $Q_{10}$ of 1.8 (6% per degree) if the exact value for a particular enzyme is unknown. Note that, as in the case of pH, $K_m$ can also change with temperature, and $Q_{10}$ refers strictly to $V_{max}$ rather than just $v$. It is not always appreciated how much difference temperature makes; sometimes students try to squeeze the last 1% of accuracy out of enzyme activity data, and then do not know to within several degrees what the temperature was! To quote an enzyme activity to within 5%, it is essential that you know the temperature of the assay solution to the nearest 0.5°C.

## 8.2   Measurement of Activity Using Stopped Methods

To measure the activity of an enzyme requires a method for determining the amount of product formed (or sometimes the amount of substrate used). This may involve separating substrates and products after the reaction; alternatively, it may be possible to measure one without interference from the other. It is convenient to divide the techniques for measuring enzyme activity into two classes: stopped methods and continuous methods. The latter will be described in the next section; continuous methods allow the progress of reaction to be monitored continuously and instantaneously, which is a great advantage. However, continuous methods have some limitations, and with many enzymes there is no suitable method available. Stopped methods can be used for an almost limitless range of enzymes, but they are generally less convenient.

In a stopped method the enzyme is incubated with its substrates for a fixed period of time, and the reaction is then stopped by one of several techniques. The product formed is then measured, and the rate assumed to be linear from time zero. One of the first things to be checked when using a stopped method

---

[1]Enzyme Nomenclature (see ref. 158).

**Figure 8.4.** Progress of reaction, and the risk of underestimating the initial activity with a stopped method due to slowing down during the reaction.

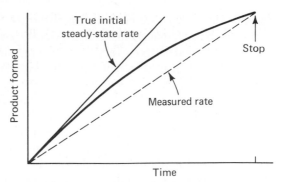

is this assumption of linearity. A single point may underestimate the initial steady state if by the time the reaction is stopped it is already slowing (Figure 8.4). Alternatively, a lag in the enzyme activity would result in an underestimate from a single point observation. By carrying out a series of incubations at different times linearity can be checked; it is especially important with crude tissue extracts, since many side-reactions are possible; the next enzyme in a metabolic sequence may be removing the product that one is trying to measure. Further discussion on measuring enzyme activity of crude tissue extracts can be found elsewhere (167).

## Incubation Conditions

For stopped methods there are no restrictions on incubation conditions; the optimum pH can be chosen—but on occasions it may be preferable to work at a pH somewhat away from optimum to shift the equilibrium to a more favorable value. Temperature is easily controlled, and the reaction mixture can be a very small volume—50 to 100 $\mu$l is common—which allows economization on reagents. The sample containing the enzyme is added at time zero and rapidly mixed in; the reaction occurs for a fixed time and is then stopped. To stabilize dilute enzyme solutions, bovine serum albumin (0.2–1 mg ml$^{-1}$) is often included in the reaction mixture.

## Stopping Methods

The enzyme reaction is stopped by a method which either causes instantaneous denaturation of the enzyme or shifts the conditions so that the enzyme is no longer active, though not necessarily denatured. The commonest methods involve acidification (usually with the protein-precipitating acids trichloracetic acid or perchloric acid) or rapid heating. The latter method has some advantages, particularly if it is undesirable to dilute the sample; but the heating must be very rapid. Very small samples in thin-walled glass (not plastic) tubes placed in a boiling water bath can reach 90°C within a second, which should be sufficient to denature most enzymes. But the rate of denaturation must be high enough to completely inactivate the enzyme within that second. Otherwise the portion of undenatured enzyme will be reacting very fast, and may yield a significant amount of additional product, especially if the activity incubation was short. Heat denaturation is best used with long incubations ($>10$ min), so that any additional product formed during the heating is relatively small.

Trichloracetic or perchloric acids at a final concentration of 3–5% stop the reaction instantly because of the very low pH ($\sim 0$) and because the enzyme denatures and is precipitated. After removing the precipitate by centrifugation, the supernatant may need neutralization before product measurement. KOH containing some $K_2CO_3$ is used; most perchloric acid is removed because the salt $KClO_4$ is sparingly soluble, especially at $0°C$. But K-trichloracetate remains in solution. Carbonate is useful in that it provides buffering at around neutrality, avoiding overshooting with possible destruction of any alkali-labile product. An internal indicator such as bromothymol blue can be used provided it does not interfere with product measurement. Some very acid-stable enzymes may not totally precipitate with these acids, and on neutralization can reactivate; this has been known to occur with adenylate kinase and potato ATPase, among others. A zero-time stopping of reaction will detect such occurrences. The zero-time experiment should in any case be carried out as it represents the true "blank" for the assay. The product being measured could be present in the enzyme sample or in the reagents, causing a non-zero blank.

Another very useful stopping method, again provided it does not interfere with subsequent product measurement, is to use sodium dodecyl sulfate at about 1% concentration. Most enzymes are rapidly denatured by this detergent, but remain in solution.

The reaction may be stopped by nondenaturing methods, provided that subsequent manipulations do not restore the conditions to allow further reaction. For example, an enzyme active only below pH 8 may be stopped by raising the pH to 9 and keeping the pH at that value. An enzyme having an absolute requirement for divalent metal ions may be stopped with an excess of ethylenediaminotetraacetate (EDTA), provided that the product can be measured or be separated from the reactant in the presence of this compound. In radiochemical assays, the reaction can be stopped by swamping with cold substrate so that very little extra labeled product is formed. Similarly, adding a large dose of a known inhibitor can be employed. There are many ways of stopping enzyme reactions which apply only to particular enzymes.

## Measurement of Product

There is no limitation on the methods that can be employed for measuring the product, provided that the stopping procedure has not introduced interfering substances. The three major methods are enzymatic, chemical, and radiochemical.

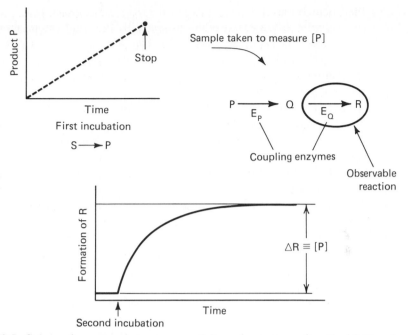

**Figure 8.5.** Scheme for measuring enzyme activity, using a stopped method followed by enzymatic detection of the products.

Enzymatic methods use purified enzymes to convert the product directly or indirectly to make an observable compound. Spectrophotometry or fluorometry are the commonly employed methods, though various other procedures (e.g., bioluminescence, gas production) are possible. If an enzymatic method is to be used, then it can usually be adapted to a continuous rather than stopped assay method. But sometimes the stopped method is more convenient, more accurate, or less expensive. Because of the specificity of enzymes, separation of product from reactants should not be needed. The product is acted on by one or more enzymes, and the reaction should preferably go to completion. Since the added enzymes do not have to "keep up with" the enzyme being assayed as in continuous methods (cf. section 8.3), much smaller amounts can be used, with the reaction progressing leisurely over a period of 10 min to an hour or so if necessary. Enzymatic analysis of metabolites requires amounts of enzyme that result in near-completion of reaction in an appropriate time. It can be shown that for 98–99% completion in 10 min, the minimum number of units of enzyme per milliliter ($V_{max}$) is given by $V_{max}/K_m = 1$ min$^{-1}$ (168), where $K_m$ is the Michaelis constant of the coupling enzyme for the relevant substrate. The process is illustrated in Figure 8.5.

As an example, consider the assay of fructose 1,6-bisphosphatase in the presence of $P_i$:

First incubation:
   Sample + fructose-1,6-bisphosphate + $P_i$ + activators, inhibitors, etc.

Stop reaction

Measure fructose 6-phosphate (P in Figure 8.5)
   $E_P$ = phosphoglucose isomerase
   $E_Q$ = glucose 6-phosphate dehydrogenase + NAD (P)$^+$

Fruct. 6-P $\rightarrow$ Gluc. 6-P $\rightarrow$ 6-P-Gluconate

$E_P$          $E_Q$

NADP$^+$   NADPH        Observe by spectrophotometry
                                      at 340 nm

Since the $K_m$ of $E_P$ for Fruct. 6-P is 0.2 mM, then one needs at least $V_{max}^P$ = 0.2 × 1 = 0.2 unit ml$^{-1}$. Because the first reaction is reversible, somewhat more enzyme should be used—say, 0.5 unit ml$^{-1}$.

If the $K_m$ of $E_Q$ for Gluc. 6-P is 0.1 mM, then one needs at least

$$V_{max}^Q = 0.1 \times 1 = 0.1 \text{ unit ml}^{-1}$$

Because the second enzyme may not be saturated with NADP$^+$, use twice as much, 0.2 unit ml$^{-1}$.

The assay for fructose bisphosphatase can be carried out in a continuous assay, but *at least* 5 times more coupling enzymes are needed (cf. section 8.3).

Chemical measurement of product $P$ cannot be generalized because of the wide variety of techniques used. If it is not possible to measure $P$ in the presence of remaining substrate $S$ (both are likely to be similar structurally), a separation procedure will be needed. Separation is always needed for radiochemical measurement and is discussed below. If no separation is needed, the chemical reagents can be added directly to the stopped solution, often without prior neutralization (if acidification was the method of stopping). To simplify procedures and eliminate centrifugation of denatured protein, some stopping method may be desirable which does not cause precipitation of protein (either during stopping or after adding the chemical reagents). As mentioned above, dodecyl sulfate denaturation leaves the protein in solution. For example, it is convenient to measure phosphate by stopping the reaction with sodium dodecyl sulfate, then adding nonprecipitating acid molybdate reagents (sulfuric acid rather than the more commonly used perchloric acid) to measure phosphate. The dodecyl sulfate does not interfere with phosphate color formation, and no protein precipitate forms.

Radiochemical methods depend on the detection of labeled product, and this must be separated completely from the labeled substrate used. This separation step is likely to be the most time-consuming and introduce the largest error into the final answer. If the product and reactant can be separated on a phase basis—e.g., precipitation of a labeled protein from soluble reactants, removal of labeled $CO_2$ from the mixture—the principle is simple, although there can be many practical problems. Otherwise some chromatographic method is needed. Paper or thin-layer chromatography, ion exchange or high-performance liquid chromatography are all possibilities. Sometimes they are quite rapid and simple, especially in an all-or-none situation where one of the product–reactant pair adsorbs totally and the other not at all. On other occasions the separation can be very tedious, especially if a very large excess of substrate is still present; the natural diffusion distribution may overlap the product position and give an erroneously high answer if the whole chromatogram is not checked (Figure 8.6).

High-performance liquid chromatography gives resolution which largely avoids the problem described above. In many cases it is not even necessary to use labeled substrate, since very small amounts can be quantitated directly by refractometry or UV absorption, and the technique is becoming widely used for difficult assays requiring high-resolution separation.

In summary, stopped methods have both advantages and disadvantages

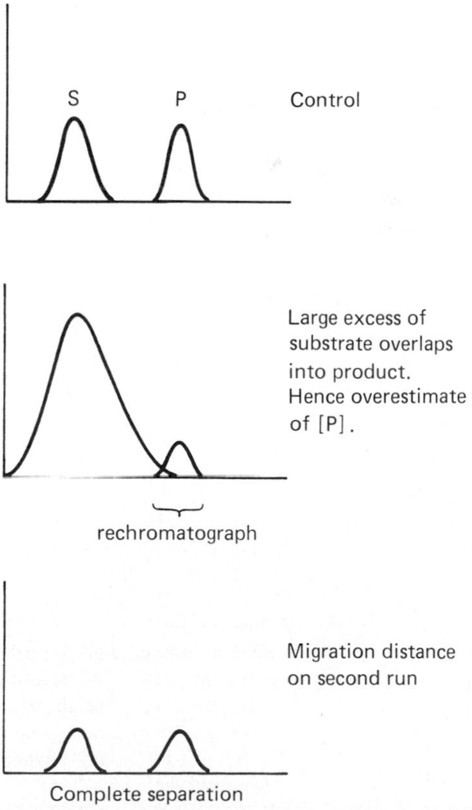

S    P         Control

Large excess of
substrate overlaps
into product.
Hence overestimate
of [P].

rechromatograph

Migration distance
on second run

**Figure 8.6.** Problems in separation
of substrate from product by
chromatography.

Complete separation

compared with continuous methods (for comparison see Table 8.3). For routine purposes the rapidity and ease of a continuous method should make it the method of choice *provided that* such a method is available and has sufficient sensitivity. Radiochemical methods must employ the stopping principle and are usually (though not always) the most sensitive.

## 8.3   Measurement of Activity Using Continuous Methods

Continuous methods imply that the progress of reaction can be observed throughout, and this is generally an instantaneous observation. A fairly large group of enzymes fits into the simple category where the formation of a product of the reaction or the loss of a substrate can be observed directly because of difference in absorption spectrum between product and substrate. Also, other results of net reaction, such as acid production or uptake, viscometric changes, or even heat production, can be followed directly. The main enzymes to which these conditions apply are dehydrogenases using NAD or NADP, many hydrolases which can utilize synthetic substrates which liberate colored or fluorescent products, and polymer endohydrolases, which cause viscosity decreases. In theory, any enzyme reaction which has an enthalpy change significantly far from zero can be followed using the sensitive calorimeters now on the market. Although not widely used, calorimetric enzyme assays could be used where no other assay is available. All of these assays could be called "uncoupled continuous assays" to distinguish them from the coupled assays described below. Some examples of uncoupled continuous assays are given in Table 8.2.

**Table 8.2.** Examples of Uncoupled Continuous Assays

| Enzyme | Reaction | Observation |
|---|---|---|
| Lactate dehydrogenase (1) | Lactate + $NAD^+$ → pyruvate + NADH + $H^+$ (high pH) | NADH formation at 340 nm |
| Lactate dehydrogenase (2) | NADH + pyruvate + $H^+$ → lactate + $NAD^+$ | NADH loss at 340 nm |
| Nonspecific phosphatases | $NO_2$◯$O-$Ⓟ→ $NO_2$◯$OH + P_i$ | Nitrophenol production at 405 nm |
| Glycosidases | Methylumbelliferyl-glucoside → sugar + methylumbelliferone | Fluorescence of methylumbelliferone |
| Xanthine oxidase | Xanthine + $O_2$ → uric acid + superoxide | Increase in absorbance at 295 nm due to uric acid |
| Endo-cellulase | Cellulose → oligo $\beta$-glucosaccharides | Loss in viscosity |
| Hexokinase | Glucose $MgATP^{2-}$ → glucose 6-$P^{2-}$ + $MgADP^- + H^+$ (high pH) | Acid production (pH-stat) |
| Chymotrypsin | $N$-$t$-BOC-L-phenylamine $p$-nitrophenyl ester → B-$t$-BOC-L-phenylalanine + $p$-nitrophenol | Nitrophenol production at 405 nm |

*Note:* In some of these examples stopped methods have greater sensitivity, since maximum absorption or fluorescence may be obtained at a pH outside the enzyme activity range.

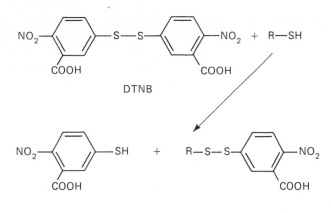

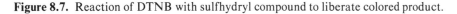

TNB

**Figure 8.7.** Reaction of DTNB with sulfhydryl compound to liberate colored product.

## Coupled Methods

The principle of coupled methods is that a product which cannot be observed directly itself is removed and reacted on either chemically or enzymically to form (directly or indirectly) a compound that can be observed. The added reagents must of course be capable of acting in the same conditions (pH, temperature, etc.) as the enzyme being measured, and must not interfere with the enzyme's activity. Most of the methods use the specificity of enzymes to select out the product and continue the mini-metabolic chain until something is formed that can be observed. But there are a few direct chemical methods in which a colored compound is formed. Of particular value is the compound 5,5'-dithiobis(2-nitrobenzoic acid) (DTNB), which reacts at neutral pH with sulfhydryl groups to liberate the yellow 5-thio-2-nitrobenzoic acid (TNB) which can be observed at 412 nm (Figure 8.7). Enzymes involving acyl-CoA, liberating free CoA, can be assayed by this method:

$$\text{Acyl-S-CoA} + \text{HX} \rightarrow \text{Acyl-X} + \text{HS-CoA}$$
$$\text{HS-CoA} + \text{DTNB} \rightarrow \text{TNB-S-CoA} + \text{TNB}$$

However, enzymes with a reactive sulfhydryl in the active center are likely to be inactivated.

Another example involves electron transfer reactions by such enzymes as hydrogenases: hydrogen uptake and passage of electrons to flavodoxins or ferredoxins can be followed using the synthetic compound methyl viologen; the absorption of this compound in the ultraviolet is largely removed on reduction:

$$\tfrac{1}{2}\text{H}_2 \rightarrow [\text{e}^-] + \text{H}^+ \rightarrow \text{ferredoxin} \rightarrow \text{methyl viologen}$$

There are several other examples of chemical coupling. But enzymic coupling is the most widely used method in continuous assays. The principles are simple, but there are many pitfalls, especially when using the method on a crude extract. The product $P$ is converted by a coupling enzyme $E_P$ to another product $Q$:

$$S \longrightarrow P \underset{E_P}{\longrightarrow} Q \underset{E_Q}{\longrightarrow} \text{etc.}$$

In the simplest case, the reaction of enzyme $E_P$ can be observed directly. For instance, consider the measurement of the enzyme triose-P isomerase:

$$\text{glyceraldehyde 3-P} \rightarrow \text{dihydroxyacetone-P}$$

The formation of the product, dihydroxyacetone-P, cannot be detected directly. It is removed by the enzyme glycerol 1-P dehydrogenase, involving NADH:

$$\text{dihydroxyacetone-P} + \text{NADH} \rightarrow \text{glycerol 1-P} + \text{NAD}^+$$

Although the product glycerol 1-P cannot be detected, every molecule that is formed results in the loss of a molecule of NADH, so the absorption at 340 nm decreases. The coupling enzyme, glycerol 1-P dehydrogenase is in this case the "indicator enzyme."

Now consider a more complex process, but involving the same enzymes. Fructose 1,6-bisphosphate aldolase is normally assayed using both triose P-isomerase as coupling enzyme ($E_P$) and glycerol 1-P dehydrogenase ($E_Q$) as indicator enzyme:

$$
\begin{array}{c}
E_Q \\
\text{dihydroxyacetone-P} \rightarrow \text{glycerol 1-P} \\
\text{fructose 1,6-bis P} \overset{\nwarrow}{\underset{\swarrow}{\big\langle}} \ \uparrow E_P \quad \text{NADH} \quad \text{NAD}^+ \\
\text{glyceraldehyde 3-P}
\end{array}
$$

It may be argued that $E_P$ is unnecessary, since the aldolase reaction already produces dihydroxyacetone-P. Apart from the advantage of doubling the sensitivity, addition of $E_P$ ensures that endogenous isomerase with the sample being assayed does not make any difference. Without an excess of triose-P isomerase, the number of molecules of NADH oxidized per fructose 1,6-bis P cleaved might vary between 1 and 2; nonlinear rates of NADH oxidation would be observed.

Still more complex, but in fact perfectly successful, is the method used for phosphofructokinase. Using the same chain of reactions as above, but this time including aldolase as a coupling enzyme, 2 NADH molecules are oxidized for every molecule of substrate converted (Figure 8.8).

These examples demonstrate how it is possible to build up a mini-metabolic sequence in order to assay an enzyme. The system should be very favorable thermodynamically so that product is removed effectively.

Sometimes more than one assay can be used; for example, phosphofructokinase has also been assayed using the general method for ADP detection, pyruvate kinase + lactate dehydrogenase (Figure 8.9).

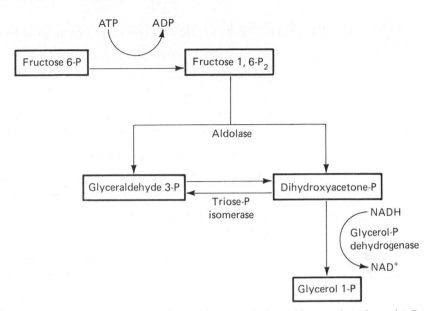

**Figure 8.8.** Principle of the coupled assay for phosphofructokinase using glycerol 1-P dehydrogenase as the indicator enzyme.

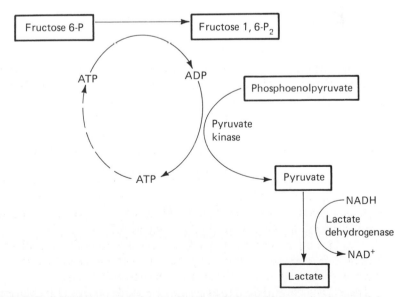

**Figure 8.9.** Principle of the coupled assay for phosphofructokinase using lactate dehydrogenase as the indicator enzyme.

Most kinases can be measured this way, replacing fructose 6-P with the relevant substrate.

The principles are clear. In practice, there are at least four important considerations:

(1) How much of the coupling enzymes are needed to ensure that they are not rate-limiting in the system?
(2) Are the coupling enzymes available in sufficient affordable amounts?
(3) Are there contaminants (particularly of the enzyme we wish to measure) in the coupling enzymes?
(4) To what extent would other enzymes present in the sample being assayed upset the system?

First consider the amounts of enzyme needed. If the answers to questions 2 and 3 are yes and no, respectively, then the answer to question 1 is to put in a large amount to be sure that it is the enzyme being measured that is limiting. Trial and error will do the same thing, but with expensive coupling enzymes it is worth knowing just how little you can get away with. Often published methods suggest quite unreasonably large amounts, which the small-budget scientist may not be able to afford. The basic principle is that a steady state should be set up in which the rate of action of the indicator enzyme is equal to that of the measured enzyme. Theoretically this cannot be achieved; it will only be approached asymptotically. So one must first define how closely we wish this condition to be achieved, and in what period of time. A typical spectrophotometric assay will run for a few minutes, and the accuracy of pipetting and of instrumentation is unlikely to be better than $\pm 2\%$. If we set our restrictions as 98% of true rate 1 min after the start of the reaction, then the amount of enzyme needed is not so very great after all.

To calculate the amounts of coupling enzymes required, one must first define a convenient rate; for NADH a $\Delta A_{340}$ of about 0.2 $min^{-1}$ is probably ideal (faster rates could mean we might run out of NADH before the 98% limit is reached). In a simple coupled system shown in Figure 8.10, the product $P$ accumulates just sufficiently to cause enzyme $E_P$ to act at the same rate.

The rate $\Delta A_{340} = 0.2$ $min^{-1}$ is equivalent to about 0.03 unit $ml^{-1}$. At the steady state $[P]$ is expected to be very small, but there must be some present in order for $E_P$ to work at all! If $[P] \ll K_m^P$ (the Michaelis constant of enzyme $E_P$ for product $P$), then the rate of $E_P$ is given by

$$v = \frac{V_{max} \cdot [P]}{K_m^P}$$

At a given time $t$ after the start of the reaction, $[P]$ is building up, so

$$\frac{d[P]}{dt} = r - \frac{V_{max} \cdot [P]}{K_m^P}$$

where $r$ = rate of enzyme being measured, a fixed value since it is saturated with its substrates.

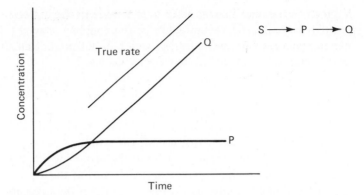

**Figure 8.10.** Concentrations of intermediate product $(P)$ and ultimate product $(Q)$ in a one-step coupled enzyme assay.

Integrating the equation:

$$\int_0^{[P]} \frac{d[P]}{r - \dfrac{V_{max} \cdot [P]}{K_m^P}} = \int_0^t dt$$

$$\ln \left[ \frac{r}{r - \dfrac{V_{max} \cdot [P]}{K_m^P}} \right] = \frac{V_{max} \cdot t}{K_m^P}$$

Now to find the conditions when $v$ is 98% of $r$,

$$\frac{V_{max} \cdot [P]}{K_m^P} = 0.98r$$

So

$$\ln \left[ \frac{r}{r - 0.98r} \right] = \frac{V_{max} \cdot t}{K_m^P} = \ln 50 = 3.9$$

If we wish $t$ to to be 1 min, then $\dfrac{V_{max}}{K_m^P} = 3.9 \text{ min}^{-1}$.

So the amount of enzyme $E_P$ needed to result in a coupling rate at 1 min that is 98% of the true rate is given by this expression. $V_{max}$, expressed in $\mu$mol ml$^{-1}$ min$^{-1}$, is the number of units of enzyme per milliliter, and $K_m^P$ is the Michaelis constant for $P$. Thus the number of units needed depends critically on $K_m^P$; units ml$^{-1}$ = 3.9 × $K_m^P$. Because there are some approximations involved (which may not be justified if $K_m^P$ is very small), we can consider the *minimum* amount of $E_P$ needed to be 5 × $K_m^P$. Note that in the discussion on the use of enzymes for measuring the *amount* (not the rate) of product formed during stopped assays (cf. section 8.2), the ammount of coupled enzyme needed was about 1 × $K_m^P$ for virtually complete reaction in 10 min. Both $V_{max}$ and

$K_m^P$ for the coupling enzyme refer to the value in the buffer conditions of the assay (not necessarily the optimum for the coupling enzyme). If the coupling enzyme involves a second substrate which is not present at saturating levels, further correction for the "effective $V_{max}$" is needed.

In the situation where two or more coupling enzymes are involved, similar calculations can be made. The amount of each needed theoretically is little greater than with one enzyme ($V_{max}/K_m \sim 10$ min$^{-1}$ for each, but the value needed for the second enzyme depends on how much of the first one is present, and vice versa). It is likely that not all the reactions in the sequence are irreversible. If so, the enzyme subsequent to the reversible reaction will be needed in greater amounts (Figure 8.11). In this situation there would be little point in putting in more of enzyme $E_P$ which catalyses the reversible reaction. But enzyme $E_Q$ must work on a lower level of $Q$, so more of it is needed.

A commonly employed measurement of kinase activity uses pyruvate kinase ($E_P$) and lactate dehydrogenase ($E_Q$) for detecting ADP production. This will be used as an example of the calculation of coupling enzyme amounts. Both of these enzymes are effectively irreversible ($\Delta G^0$ large negative), so $V_{max}/K_m$ values of 5 min$^{-1}$ should be sufficient. For pyruvate kinase, the $K_m$ concerned is for ADP. There is some confusion in the literature as to whether the true substrate is ADP$^{3-}$ or MgADP$^-$, and the amount of either will depend on the presence of free magnesium ions which are otherwise essential for the reaction. A value of 0.2 mM is a reasonable figure to use for the $K_m$. Phosphoenol pyruvate, the other substrate, will normally be used at 0.2–0.5 mM concentration, which is close to saturation. Potassium ions are essential, and for full reactivity should be present at 0.1–0.15 M. If the kinase being measured cannot be assayed in the presence of such high concentration of K$^+$, then an appropriate adjustment of the $V_{max}$ for pyruvate kinase must be made. The amount of enzyme should be at least 1 unit ml$^{-1}$ ($5 \times K_m^{ADP}$). The quoted value for a commercial sample of pyruvate kinase will inevitably refer to optimum conditions, so doubling the figure to 2 (optimum) units ml$^{-1}$ is advisable. When measuring an enzyme at pH 8.0, where pyruvate kinase has relatively low activity, use at least 5 times more than the theoretical figure, i.e., 5 units ml$^{-1}$.

The amount of lactate dehydrogenase needed will depend on its origin. The $K_m$ concerned is that for pyruvate, and skeletal muscle enzymes have higher $K_m$ values than heart muscle enzymes. On the other hand, the skeletal muscle enzymes' specific activities are also somewhat higher. If $K_m$ for pyruvate = 0.5 mM, then 2.5 units ml$^{-1}$ lactate dehydrogenase is the minimum. However, it may not be completely saturated with its other substrate NADH, so 5 units ml$^{-1}$ should be used. Thus more units of lactate dehydrogenase may be needed than of pyruvate kinase, not because it is the second enzyme, but because it has a lower affinity for its substrate.

The actual concentration of the intermediates ADP and pyruvate can be calculated; if $V_{max}/K_m$ values of 5 have been used, and the actual rate $v = 0.03$ unit, then in the steady state [ADP] $= K_m^{ADP} \times 0.03/5 = 0.0012$ mM and [Pyr] $= K_m^{Pyr} \times 0.03/5 = 0.003$ mM. In each case these satisfy the

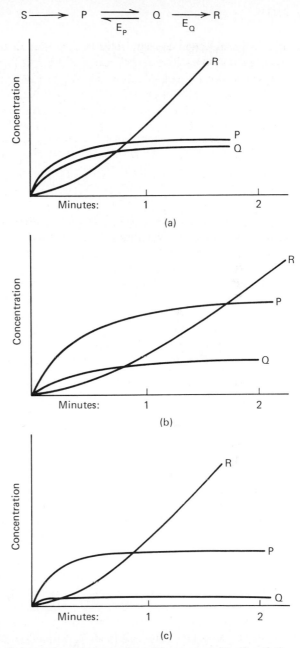

**Figure 8.11.** Two-step coupled reaction where product $P$ is converted to $Q$ and then to $R$ by enzymes $E_P$, $E_Q$. Concentrations of $P$, $Q$, and $R$ (a) when $V_{max}/K_m = 5$ min$^{-1}$ for both $E_P$ and $E_Q$, and $P \rightarrow Q$ is essentially irreversible. (b) $V_{max}/K_m = 5$ min$^{-1}$ for both $E_P$ and $E_Q$, and $P \rightarrow Q$ is reversible with equilibrium favoring $P$. Note that rate of formation of $R$ is still increasing at 2 min. (c) as (b), but $V_{max}/K_m = 25$ min$^{-1}$ for $E_Q$. Rate of formation of $R$ reaches a steady state by 1 min, with $[Q]$ much lower than in (b).

original condition that [substrate] $\ll K_m$, so that use of the simplified expression

$$v = \frac{V_{max} \cdot [S]}{K_m}$$

is justified.

Should it be required that the coupling system reaches 98% of the true rate in a shorter time, then the amounts of coupling enzymes needed must be increased by the appropriate factor, e.g., for 10 sec rather than 1 min, multiply the above enzyme concentrations by 6. For further analysis of coupled enzymic assays, see Ref. 169.

Having done these calculations and estimated the number of assays projected, one must ask whether sufficient amounts of these coupling enzymes are available and affordable? In the case described above, where in a 1-ml-volume assay 2 units of pyruvate kinase and 5 of lactate dehydrogenase are needed, the cost is negligible. Each is available for a fraction of a cent per unit. On the other hand, lower specific activity enzymes cost much more per unit, and high $K_m$ values may make the unit requirement greater, so that some perfectly good methods are prohibitively expensive on a routine basis. Also included in this question about availability (which should really be asked right at the beginning) is the more direct question not just of cost, but whether they are available at all commercially, and if not, can one make them in the laboratory? Consider the enzymes of the Entner–Doudoroff pathway. To measure the enzyme 6-phosphogluconate dehydratase, the next enzyme on the pathway, 3-keto-2-deoxy-6-phosphogluconate (KDPG) aldolase is needed. This is not available commercially, and has to be purified. To assay it, the substrate is needed. This is also not available commercially—but that is another story! Once one has a stock of the aldolase, one can measure the dehydratase readily:

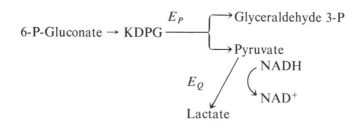

where $E_P$ = KDPG aldolase, and $E_Q$ = lactate dehydrogenase.

A third question concerns contaminants. If the coupling enzymes contain any of the enzyme being assayed, a rate will be observed even in supposedly blank samples. If this is not great, then a simple correction for it with each assay is all that is needed (Figure 8.12). Contamination of this sort is obvious and easy to correct for, although it may limit the sensitivity of the assay. Other contaminants are not so obvious, and one is not always aware of interference until too late.

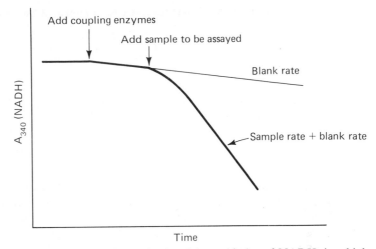

Figure 8.12. Progress of a coupled reaction involving oxidation of NADH, in which there is a "blank rate" due to contamination in the coupling enzymes.

Assays of crude extracts commonly pose problems because in addition to the desired enzyme a whole host of others are present which might cause upsets. The problems divide into two categories: first, the crude extract may contain sufficient of other substrates (as well as enzymes) to allow other detectable reactions to be occurring; second, other enzymes may be interfering with the coupling system being used. Problems mainly occur when the enzyme being measured is relatively weak in activity, since this means that larger amounts of sample must be added, increasing the probability of detectable interference. For instance, if a kinase is being measured as ADP formation by the pyruvate kinase/lactate dehydrogenase method described above, addition of sufficient glucose in the sample (likely in liver extracts) will activate hexokinase as well as the kinase being measured. NADH oxidation may occur due to the presence of an "NADH oxidase" system. NADPH formation can be decreased by the presence of glutathione reductase plus oxidized glutathione. There are many possibilities, and if interference is suspected, the best way to find out what is going on is to ask, "What am I looking at?" The answer is (usually) absorption at 340 nm. The next questions are: "Could it be a turbidity change, or is it due to NADH?" "What else could cause a decrease/increase in NADH?" and so on. Problems such as removal of substrate by another enzyme or removal of product by a side reaction may be overcome by various blank determinations or possibly introduction of inhibitors. This is discussed in more detail elsewhere (167).

Having described the principles of both stopped and continuous assays, a comparison of the relative advantages and disadvantages of continuous compared with stopped assays can be presented. These have already been pointed out earlier, and Table 8.3 lists them systematically. Finally, consider what can be called a "continuous-stopped" method! Using an automated assay system it

**Table 8.3.** Comparison of Advantages and Disadvantages of Stopped and Continuous Enzyme Assays

| Stopped assays | Continuous assays |
|---|---|
| No limitation on conditions | Conditions may be limited by requirement of coupling system |
| Very small volume saves on expensive reagents | Usually requires larger amounts of reagents, though microcells result in saving |
| High sensitivity for radiochemical methods | Sensitivity usually low compared with radiochemical methods |
| Single point determination may hide irregular activity | Continuous observation detects nonlinearity |
| Result not instantaneous | Instantaneous result |
| Multi-manipulation more time-consuming | Simple manipulation |
| No coupling enzymes required in assay (but may be needed to determine product) | Coupling enzymes needed may be expensive and/or contain contaminants |
| Product accumulation may cause inhibition[a] | Product may be removed by coupling system |

[a] A nonobservable coupling system can be included to remove products.

is possible to stop a sample of the reaction mixture and then measure the products, so that a continuous trace is acquired. The example shown in Figure 8.13 demonstrates the application of this method to a phosphate-liberating enzyme. It is possible to carry out many other assays in this way where otherwise a stopped method would have been necessary.

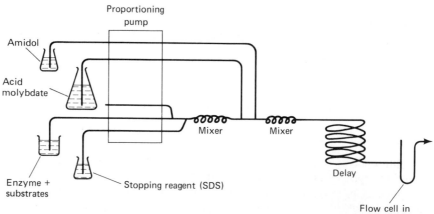

**Figure 8.13.** Flow scheme for a "continuous-stopped" assay, in which the product of the enzyme, phosphate, is measured. A sample of the reacting mixture is continuously taken up, reacted with phosphate reagents, and the progress of phosphate formation recorded.

# 8.4  Practical Points in Enzyme Activity Determination

The theory and basic practice of enzyme activity measurement have been broadly described above, including many "*do*s and *don't*s." This short section amplifies the practical laboratory aspect of actually carrying out the measurements.

A typical day during an enzyme preparation may involve only one or two assays, but more probably it will be 10 or 20; monitoring a column effluent may need even more. Most assay mixtures involve many components—buffers, salts, substrates, cofactors, etc.—and it is not only inconvenient to mix up each reaction mixture separately, it is also bad practice since slight variations and errors will make each mixture somewhat different. The best method is to estimate how much will be needed in the day, and make up the whole mixture in bulk with the exception of one substrate, to be added at the last moment to each assay. Some reagents can be mixed and kept (refrigerated) for weeks— for instance, simple buffer and salt solutions. Thus "Reagent A" might include all of the buffers, salts and perhaps cofactors, needing only one or two other components to complete the mixture. Many biochemical reagents are quite stable in solution unfrozen, provided that bacterial or fungal activity is prevented. One drop of 20% sodium azide for every 20 ml usually gives sufficient protection; many stock solutions can be kept in the refrigerator for months like this. Others may be chemically labile and so need mixing fresh each day. Sodium azide rarely affects enzymes, and is in any case diluted out to low levels if the stock solutions of reagents are concentrated. As an example, below is a protocol for mixing up a hexokinase assay:

*To make 20 ml of "Reagent A" for hexokinase assay*:
   Take 2 ml of stock buffer pH 8.0 (see below)
   Add: 1-2 drops of 5% bovine serum albumin
         0.4 ml of stock 10 mM $NADP^+$
         0.4 ml of stock 50 mM ATP
         10 $\mu$l of glucose 6-P dehydrogenase, 1000 units $ml^{-1}$ suspension
            in ammonium sulfate
   Make to 20 ml with water.

   Stock buffer contains 0.3 M Tris, 0.5 M KCl, 30 mM $MgSO_4$, 1 mM
      EDTA (+ azide), pH to 8.0 with HCl.

*Assay procedure*:
   Take 1 ml of Reagent A in cuvette
   Add 20 $\mu$l of stock M glucose ("Reagent B")
   Check $A_{340}$ for any blank rate

   Add sample (0.2 $\rightarrow$ 10 $\mu$l), measure increase in $A_{340}$.

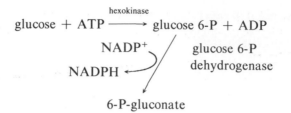

$$\text{6-P-gluconate}$$

(*N.B.* Glucose 6-P dehydrogenase from *Leuconostoc mesenteroides* can use NAD as an alternative to NADP, which can be more economical.)

If glucose is placed in the "Reagent A" mixture and ATP used as "Reagent B," glucose 6-phosphate dehydrogenase slowly oxidizes glucose; by the end of the day the absorption of the solution has increased. This is avoided by putting the ATP in Reagent A mixture and adding glucose at the last minute. Note that unless dealing with very labile compounds, Reagent A should *not* be kept on ice; the temperature should be close to that required in the assay to minimize temperature equilibration time in the cuvette. This does not apply to stopped assays, since a short preincubation is easily done. Temperature control in spectrophotometers can be arranged using a cuvette holder with circulating water; although this will control the temperature accurately, there are two points to remember. First, the actual temperature in an equilibrated cuvette may be somewhat different from the set temperature on the circulating water thermostat because of heat losses during transmission. Second, it could take up to 10 min for a cold sample in a cuvette to reach within 0.5°C of the steady-state temperature; this is obviously inconvenient, so preincubation is desirable. Because of these points it is usually more convenient for routine work to carry out assays at room temperature, noting the actual temperature in the cuvette, and correcting to a standard temperature using an appropriate factor. Provided that the temperature is within 5°C of the standard 25°C, a factor of 6% per degree is a good average value for many enzymes (cf. section 8.1). It does not take long to find the correct factor for one particular enzyme; simply assay the same sample over the range of temperatures likely to be used.

With a mixture containing all the ingredients, preincubated to a suitable temperature, the final step is to add the enzyme. With very-low-activity samples the volume needed may be 20% or more of the final mixed volume, but concentrated samples of very active enzymes may need diluting as much as a million-fold into the assay mixture. Microliter syringes can deliver as little as 0.5 $\mu$l with 1–2% accuracy, but smaller volumes than this inevitably involve larger errors. A range of microliter syringes from total capacity of 1 $\mu$l up to 100 $\mu$l should be on hand so that the volume desired can be added with optimum accuracy. If less than 0.5 $\mu$l is needed, routine assays can make use of the smallest syringes down to 0.1 $\mu$l or even less bearing in mind that the accuracy is low. But reliable, reportable measurements must be done using at least half of the deliverable volume. This means that some dilution of the sample is usually necessary. This dilution should be done with maximum accuracy since it is an additional step which can introduce further errors. One way to dilute samples is to use 100 μl plus 0.9, 1.9, 4.9, or 9.9 ml of dilution buffer from a graduated pipette to give 10×, 20×, 50×, or 100× dilutions, respectively. The dilution buffer's composition is important. The protein concentration of the sample could already be quite low, and to avoid inactivation/adsorption losses due to low protein concentration, a neutral protein must be added. Bovine serum albumin is normally used, and a concentration of 1% in the dilution buffer is not unreasonable. The buffer itself should be optimal for stability of the enzyme; if these conditions are not known, then a pH 7 buffer, e.g., phosphate, containing magnesium ions (1–2 mM), and a small amount of EDTA (0.1 mM) is a good starting point. If the enzyme diluting buffer contains 5 mM sodium azide as well, it also will keep for months.

# 8.5  Methods for Measuring Protein Concentration

All enzymes are proteins, and the total amount of protein present at each step in a fractionation scheme is useful to know. Nevertheless, it is much more important to follow the activity of the enzyme; the protein can be measured later if required. The times when accurate protein estimation is needed are:

(1)  When a fractionation step is critically dependent on the protein concentration
(2)  When it is necessary to know whether a particular step has really removed much unwanted protein
(3)  To test the specific activity of the penultimate or final preparation
(4)  To present the progress of purification as a final report

At other times it can be useful to know roughly how things are progressing as far as protein content is concerned, but often one gets sufficient information from experience—just how large a particular precipitate is, how big the peak on the recorder was, etc. There is no real need to know accurately what the protein content is at all stages.

This introductory paragraph was put in at the start of this section to indicate that it is possible to carry out routine enzyme preparations without *ever* measuring protein content, or perhaps just measuring it in the final preparation. One should not get held up during a fractionation procedure while measuring protein—small samples should be put aside and measurement done later unless it is critical for the procedure to know the protein concentration.

That said, some of the available methods for measuring protein are now described. Several which require special apparatus (e.g., Kjeldahl nitrogen, refractometry) will be omitted. The most widely used methods are:

(1)  Biuret—alkaline copper reagent

(2)  Lowry—Folin–Ciocalteau reagent
(3)  UV absorption at 280 nm (aromatic band) or 205–220 nm (peptide band)
(4)  Dye binding

Each of these has particular advantages and disadvantages, but it is worth noting that *no* method, with the exception of dry weight determination, gives an answer that is unambiguously correct to within a few percent unless it has been calibrated against a protein solution of identical composition. In practice, this means that really accurate measurements can only be made for pure proteins, after those pure proteins have been standardized for that method against a dry weight determination. If the protein contains nonproteinaceous prosthetic groups, carbohydrate, or nucleic acid, dry weight determination gives a value for the whole molecule rather than just the protein portion. The principles of the four methods are as follows.

## Biuret Reaction

This involves a strongly alkaline copper reagent which produces a purple coloration with protein. The principle reason why it is not widely used is its low sensitivity; several milligrams of sample must be sacrificed for a reliable measurement. The method gives an accurate measurement since there is little variation in color yield from protein to protein. This is because the copper reagent reacts with the peptide chain rather than side groups. Ammonia interferes by complexing with copper, so ammonium sulfate fractions do not give accurate results. A greater sensitivity of the biuret reaction is achieved by observing the copper–protein complex not at 540 nm, but at 310 nm in the near-ultraviolet (170). Unfortunately, this method needs several blanks to subtract spurious absorption at this wavelength, so it has not been widely employed.

## Lowry Method

The most cited reference in biochemistry is that of Lowry, Rosebrough, Farr, and Randall (171) on a new method for protein estimation. A combination of the copper reaction used in the biuret method and the reaction of the Folin–Ciocalteau reagent (172) with phenols such as tyrosine in proteins was found to give a strong dark blue color with proteins. The original paper surveyed ranges of concentrations of reagents to determine the ideal values. Since then many modifications have been reported, mainly to avoid interference by specific components. Unfortunately, many of the compounds used in enzyme purification interfere with the reaction, and different modifications are required to overcome each problem. But since it is a sensitive method giving a good color with 0.1 mg ml$^{-1}$ of protein or less, interfering compounds are often diluted out to levels where their effect is insignificant. An extensive review of the method and the many variations on it has been published recently (173).

## UV Absorption

This has been employed for measuring protein ever since Warburg and Christian reported a method based on 260-nm and 280-nm absorption which would correct for nucleic acid and nucleotide content (174). This correction is very important with crude extracts, but becomes less relevant as protein purification progresses and interfering compounds are removed; a simple 280-nm reading is then sufficient. Proteins absorb at 280 nm solely because of tyrosine and tryptophan residues (unless they also contain UV-absorbing prosthetic groups). Since the content of these two amino acids varies enormously, the extinction coefficient, usually expressed either as $E_{280}^{1\%}$ or $E_{280}^{1\ \text{mg/ml}}$, varies considerably. Most proteins fall in the range (for 1 mg/ml) 0.4–1.5, but extremes include some parvalbumins and related $Ca^{2+}$-binding proteins (0.0) at one end and lysozyme (2.65) at the other end. Absorption at 280 nm gives only a rough idea of the actual protein content, except with pure proteins. If the extinction coefficient for the pure protein is accurately known (and it should be standardized against dry weight, or at least against 205 nm absorption—see below), then the reading at 280 nm provides an accurate measure of the protein content—indeed, it is the most accurate method for a pure protein, since no manipulation other than appropriate dilution is required. Of course, the solvent must not absorb at 280 nm, or if it does, an accurate blank must be taken. The method is nondestructive, so, although a milligram or so is required for accurate determination, this can be returned to the bulk sample.

Absorption in the far-UV around the peptide band is used as a far more sensitive method, much less affected by the protein amino acid composition (175,176). Because of far-UV-light absorption by oxygen it is not possible to measure at the peak of the peptide absorption at 192 nm using routine spectrophotometers. But measurements on the side of the band give sufficiently accurate results provided that the absorbance is kept down to no more than 0.5. The extinction coefficients for proteins (1 mg per ml solution) are approximately:

|           |    |
|-----------|----|
| 220 nm:   | 11 |
| 215 nm:   | 15 |
| 210 nm:   | 20 |
| 206 nm:   | 29 |
| 205 nm:   | 31 |
| 200 nm:   | 45 |
| Peak at $\sim$192 nm: | 60 |

The value at 206 nm is included, as this is the wavelength employed in far-UV monitors such as the LKB-Uvicord 3.

These values vary from protein to protein because aromatic and some other residues also add variable contributions in this range, and the secondary structure ($\alpha$ helix, $\beta$ sheet, etc.) has some influence on the shape and exact position of the peptide absorption peak. By analyzing possible variations, it can be shown that the most consistent value is obtained around 205 nm (175). The extinction coefficients of proteins nearly all fall in the range 28.5–33 at this wavelength. By making a correction for the tyrosine and tryptophan content (by determining the 280-nm absorption also), the extinction coefficient at 205 nm can be predicted with no more than 2% error (177). The formula used is:

$$E_{205}^{1\text{ mg/ml}} = 27.0 + 120 \times \frac{A_{280}}{A_{205}}$$

Since 205 nm is close to the useful limits of routine spectrophotometers, measurements at 205 nm require very clean cuvettes, a relatively new deuterium lamp, and buffers with minimal absorption. Most salts absorb at this wavelength; about the only common anions that do not are sulfate and perchlorate. One can use a weak phosphate buffer (5 mM, pH 7.0) in the presence of 50 mM sodium sulfate—this salt helps to avoid adsorption of protein to the cuvette. The most convenient method is to zero the spectrophotometer with 3 ml of buffer, then mix in 1–10 $\mu$l of protein sample (containing between 2 and 50 mg/ml), and note the increase in absorption. If the buffer composition of the protein sample is known, then the same amount of that buffer can be used to determine a blank. Virtually all commonly used buffers absorb quite strongly at 205 nm, but because the dilution is of the order of a thousand-fold, the actual effect is usually very small. For more dilute protein solutions the buffer blank is relatively larger, so use of the method may be impractical. If less accurate estimates ($\pm 10\%$) are sufficient, then use of $E_{205}^{1\text{ mg/ml}} = 31$ is an adequate compromise. For use of the 280/205 absorption method, the 280-nm determination must be done on a more concentrated sample and the appropriate factor used in the equation.

## Dye Binding

This method (178,179) has become very popular recently because it is very sensitive, fast, and at least as accurate as the Lowry method. The main problems are technical; the dye adsorbs to glassware and to cuvettes (and skin!), making it a rather unpleasant reagent. Coomassie Blue G-250 dissolved in perchloric acid turns a red-brown color, but when it binds to protein in this solvent the blue color is restored. The procedure is simply to add a sample of protein to the reagent and measure the blue color at 595 nm. Use of disposable cuvettes is recommended. However, adsorbed blue dye can quickly be removed with a rinse in dilute sodium dodecyl sulfate. The color developed depends on the composition of the protein to a limited extent, so accurate determinations require standardization with an appropriate protein mixture.

This brief summary of some of the more popular methods of protein measurement gives an idea of what is possible. Each has its advantages and disadvantages, summarized in Table 8.4. The objective of an enzyme purification is to get rid of as much protein as possible while retaining enzyme activity, and

**Table 8.4.** Relative Advantages and Disadvantages of Some Commonly Employed Methods of Protein Determination

| Method | Amount of protein needed (mg) | Destructive? | Variation of response with amino acid composition | Comments |
|---|---|---|---|---|
| Biuret | 0.5–5 | yes | low | Caustic reagent; $NH_4^+$ interference; rapid color |
| Lowry | 0.05–0.5 | yes | moderate | Slow color development |
| Absorbance at 280 nm | 0.05–2 | no | large | Interference by UV-absorbing materials; instantaneous |
| Absorbance at 205 nm | 0.01–0.05 | no | low | Interference by UV-absorbing materials; instantaneous |
| Dye binding | 0.01–0.05 | yes | moderate | Acid reagent: color adsorbs to glassware; rapid color formation |

the success of each step depends both on the retention of activity and the extent of improvement in specific activity, expressed in activity units per milligram protein.

Detailed recipes for some of the reagents used for protein estimation are given in Appendix B.

# Chapter 9
# Analysis for Purity; Crystallization

## 9.1 Electrophoretic Analysis

Resolution of separate components of a protein mixture is achieved most clearly using an electrophoretic method. As discussed in section 5.2, *preparative* electrophoresis has major problems despite its theoretical potential, and is consequently not often used. But in an *analytical* mode, electrophoresis is a most widely used method; indeed, it is almost obligatory to characterize a purified protein preparation by an electrophoretic technique. Analytical electrophoresis in a gel system requires only 5–25 μg protein; this is rarely a significant proportion of what is available. Before gel systems were developed, electrophoretic analysis was carried out in the Tiselius free-boundary apparatus, requiring tens of mgs of protein. This did not resolve closely similar proteins, and analyzing a single sample required a great deal of effort and attention. Paper and other cellulose-based supports were introduced for zone analytical electrophoresis, which eliminated two of the disadvantages of the Tiselius apparatus; only small amounts of protein were needed, and the technique was easy, requiring only simple equipment. But resolution is scarcely any better even on the superior modern cellulose acetate strips, because separation is based only on a rough charge/size ratio; many proteins move together as a single peak (cf. section 5.2).

When Smithies (128) used starch gel as a support medium, an immediate improvement in resolution was achieved. As a result of the small pore sizes in the gel (compared with paper, agar gels, and other materials that had been used up till then), large molecules were retarded. This is a result of the apparent increase of viscosity; effective viscosity becomes dependent on molecular size. Moreover, diffusion is lessened, so very sharp zones could be found even after overnight electrophoresis. A few years later, Ornstein (129) introduced a

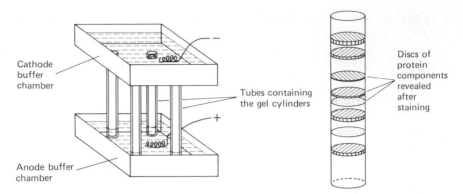

**Figure 9.1.** Disc gel electrophoresis system. Each protein component resolves as a disc in the tube of polyacrylamide gel.

synthetic gel medium, cross-linked polyacrylamide, which is more controllable than starch and has some other desirable advantages. Polyacrylamide gel electrophoresis was originally associated with "disc gel electrophoresis," because the system was promoted in which samples are run in individual tubes of gel, resolving into "discs" of protein zones (Figure 9.1). Although still widely used, disc gel electrophoresis is being replaced by slab gels, on which several samples can be run simultaneously with direct comparison of mobility. Thin-slab gel electrophoresis, using polyacrylamide, has been developed extensively over the past few years, and the technical problems associated with pouring and with sample application have been resolved, so that the system is as simple as, but more powerful than, disc gels. A variety of commercial apparatuses are available, using gel thicknesses down to less than 1 mm (Figure 9.2). One advantage of a thin gel is that less heat is produced per square centimeter of gel surface for a given applied voltage. Also, during staining and destaining of the protein bands diffusion of dye is more rapid into the thin gel slab.

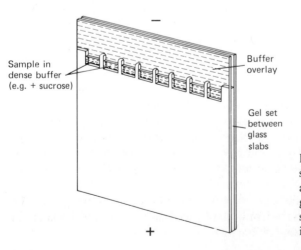

**Figure 9.2.** Modern thin-gel slab apparatus. Samples are applied in the wells set in the gel using a "comb." Side-by-side comparison of samples is possible in slab systems.

There are at least five distinct electrophoretic procedures using polyacrylamide gel as the medium:

## Simple Electrophoresis

This involves running the sample in a buffer at a pH where the proteins remain stable and in their native form. This method was the original procedure, making use both of differences in charges between proteins and their different sizes. The buffer chosen depends somewhat on the nature of the proteins, but generally it is slightly alkaline, in the pH range 8–9, where most proteins are negatively charged and so move toward the anode. The anode is normally at the bottom of the gel. It should be noted that there is no provision for proteins that move in the other direction in these systems (Figures 9.1, 9.2); basic proteins disappear into the cathode buffer. If the bulk of the proteins being observed are known to be basic, then a buffer of somewhat lower pH can be employed, and the system operated with the cathode at the bottom of the gel.

The concentration of the polyacrylamide can be varied over a wide range according to the size of molecules being separated. Two variations are possible: variation in total acrylamide content and variation in cross-linker percentage ($N,N'$-methylene bisacrylamide). Each achieves roughly the same objective; an increase in either reduces the pore size and so slows up larger molecules more. For very large proteins, up to $10^6$ daltons, an open, rather difficult-to-handle gel of about 3–4% acrylamide and 0.1% bisacrylamide can be made. For very small proteins around $10^4$ daltons, 15% acrylamide and/or up to 1% bisacrylamide may be used. A high percentage of cross-linker tends to make the gels more opaque; 1 part of cross-linker to between 20 and 50 of monomer is the normal range. Usually gels of 7–10% acrylamide are used. Polymerization is initiated with freshly dissolved ammonium persulfate (1.5–2 mM) together with a free radical scavenger, TEMED (0.05–0.1% v/v, $N,N,N',N'$-tetramethylethylenediamine). Alternatively, photopolymerization using riboflavin may be used; it leaves less harmful residue in the gel. Gelation should occur within 30 min at room temperature. Further details of suitable buffers and other parts of the system are available from apparatus manufacturers.

Simple electrophoresis can also be carried out in starch gel as the medium. The quality of the starch and its behavior in forming gels can be rather variable, but before polyacrylamide systems were widely used, starch was very popular. It is still particularly useful when detecting enzymic activity after electrophoresis; a starch gel slab can easily be sliced and stained on one-half for protein, the other half for enzyme activity (see Figure 9.6). Because starch gels are opaque, protein staining is a surface stain, so a larger protein amount is needed than with transparent gels in which the whole depth of the protein zones is stained.

Simple electrophoresis in agarose gels does not have the resolving power of smaller-pore gels, so an analysis of purity based on homogeneity in such an electrophoretic system is less convincing.

## Urea Gels

The second method is particularly useful for proteins that are insoluble at the low ionic strength needed in electrophoresis. This involves electrophoresis in the denatured state in the presence of between 6 and 8 M urea. The sample is dissolved in urea to completely denature the components first. Generally a little mercaptoethanol is included to prevent disulfide bond formation between poly-peptide chains, and each protein is separated according to charge and to *subunit* size. The limitations of these gels are similar to those run in the absence of urea.

Urea-containing starch gels are also satisfactory and can be easier to handle than polyacrylamide-urea gels. Although transparent in the presence of the urea, during staining and washing the urea is lost and starch gel becomes opaque, so again only the surface-stained protein is visible.

## SDS Gels

This third method can also be used for proteins insoluble at low ionic strength; it is presently the most commonly used method for all types of proteins. It involves denaturing the proteins with the detergent sodium dodecyl sulfate (Figure 9.3). Commonly known as PAGE-SDS (*polyacrylamide gel electrophoresis in sodium dodecyl sulfate*), this high-resolution method has won most protein chemists over, partly because one can describe the bands not just in terms of their relative mobility to each other, but also in terms of their molecular size, for PAGE-SDS separates polypeptide chains according to size. Dodecyl sulfate binds strongly to proteins (180), so that only 0.1% dodecyl sulfate is sufficient to saturate the polypeptide chains, with approximately 1 detergent molecule per 2 amino acid residues. Each dodecyl sulfate carries a negative charge, so a typical polypeptide of molecular weight of 40,000 acquires about 180 negative charges—far in excess of any net charge that might exist (at neutral pH) on the polypeptide chain originally. Consequently the charge/size ratio is virtually identical for all proteins, and separation can occur only as a result of the molecular sieving through the pore of the gel. Despite the fact that the potential of separation of proteins of identical size is not possible in this system, it nevertheless appears to give the sharpest overall resolution and cleanest zones of any method. What is more, by comparison with a mixture of stan-

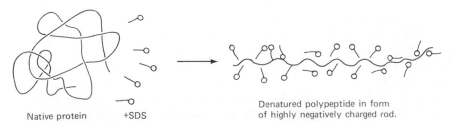

Native protein        +SDS               Denatured polypeptide in form
                                          of highly negatively charged rod.

**Figure 9.3.** The action of dodecyl sulfate in denaturing proteins.

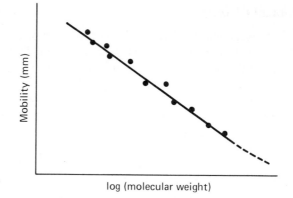

**Figure 9.4.** Plot of mobility in polyacrylamide–sodium dodecyl sulfate gel electrophoresis versus log of the molecular weight for a range of purified proteins. [From Weber and Osborn (132).]

dard polypeptides of known molecular weight, the whole gel can be calibrated in terms of mobility against size. It is found that a linear plot over a substantial range can be obtained if mobility is plotted against log (molecular weight) (Figure 9.4) (132).

For analyzing the purity of an enzyme preparation, one hopes that there will only be one component; but if there is more than one (and the enzyme itself can be identified), then the possible nature of the impurities may be guessed at. The guess may not be correct, but it is not uncommon to identify an impurity from its size. For example, phosphoglycerate kinase from silver beet leaves was purified, and after passage through a gel filtration column it still contained a substantial amount of impurity (181). Since the molecular weight of phosphoglycerate kinases is around 45,000 (a monomer), and the impurity band was at 26,000, the conclusion was that the impurity was a dimer, MW 52,000, which did not separate well on gel filtration. A well-known enzyme of carbohydrate metabolism, triose phosphate isomerase, is a dimer of $2 \times 26,500$ (yeast or muscle), and on enzymically testing the preparation this indeed was the impurity. Complete separation was achieved making use of phosphoglycerate kinase's affinity for the dye ligand Blue-Sephadex.

Clearly an impurity with the same subunit size as the desired product will not be detected in this system; however, it can resolve components differing by as little as 1% in molecular weight in optimum conditions. In fact, the mobility/molecular weight ratio is not accurate to within 1%, so strictly speaking one should speak of resolving components with mobility differences corresponding to 1% in molecular weight; their actual molecular weights could be identical. It should be noted that many proteins have more than one subunit type, so multiple bands are not necessarily an indication of the presence of impurities. If this is suspected, then there should be a rational relationship between the intensities of the multiple components. For example, a protein of subunit structure $\alpha_2\beta_2$, where $\alpha$ is of 50 kD (kilodaltons, or molecular weight $\times 10^{-3}$) and $\beta$ is of 30 kD, should show two bands for which the 30-kD stains with only 60% of the intensity of the 50 kD one (as it has only 60% as much protein in it). Some comments on estimating relative intensities are given later.

## Gradient Gels

The fourth method to be described is also widely used; it resembles the sodium dodecyl sulfate method in that it separates protein according to size only, not charge differences (182). A slab of polyacrylamide is poured in which the acrylamide concentration varies from a high value at the bottom (usually about 30%) to only 3% at the top. The buffer is a high-pH one, so that most proteins migrate into the gel toward the anode at the bottom. Electrophoresis is continued until all proteins have reached a thickness of gel which prevents them from moving any further; the small molecules may reach 25% acrylamide, large ones remaining near the top. As with the dodecyl sulfate system, molecular size (this time of the native protein, not subunits) can be determined by comparison with a standard mixture. The gradient gel system has a similarity with the next, fifth, method; it is run until no further movement occurs, and a "focusing" or concentration of diffuse protein zones can occur, resulting in very fine resolution.

## Isoelectric Focusing

The principles of isoelectric focusing were described in section 5.3. Analytical isoelectric focusing in slab gels of polyacrylamide or agarose is a popular, very-high-resolution procedure. Components are separated according to their isoelectric points, and bands are sharpened by the "focusing" effect. The gel in this case serves only as a stabilizing medium and should preferably not slow down large molecules. Since molecules eventually have to reach their isoelectric point, viscous barriers to mobility should be avoided. With polyacrylamide an open-pore gel is indicated; however, these can be difficult to handle, and sometimes polymerization is uneven, giving a nonhomogenous product. If the proteins of concern are all fairly small (e.g., <100 kD) then a 5–6% polyacrylamide gel will be satisfactory. Otherwise agarose, which sets to a firm gel at only 2% w/v, with large pores, is more appropriate.

Several samples can be applied to the one gel, and it does not matter where they are applied, since each component will eventually find its way to its appropriate spot in the pH gradient. An ampholyte concentration of between 2 and 4% is usually advised. Pre-set gels containing ampholyte can be purchased; these give consistent results with predictable pH gradients when equilibrium has been reached. Because the conductivity of ampholytes tends to zero as the focusing progresses, it is possible to apply high potentials, 100–200 V cm$^{-1}$ without excessive heating. A high-potential gradient ensures rapid focusing; modern systems can be run to completion in a couple of hours.

One problem with isoelectric focusing is that on occasion proteins may associate with ampholyte components, acquiring a combined isoelectric point different from that of the free protein. This can cause multiple minor bands, since ampholytes themselves are not a continuum of species but occur as discrete components. Thus, although a single band on isoelectric focusing is a very good

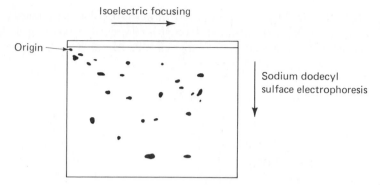

**Figure 9.5.** Two-dimensional gel electrophoretic analysis.

indication of homogeneity, multiple minor components may not reflect true heterogeneity of the protein sample.

## Two-Dimensional Systems

Two-dimensional gel analysis has become a widely used technique. Commonly a sample strip containing proteins that have been separated by a procedure such as isoelectric focusing is applied along the edge of a gel, and the proteins electrophoresed perpendicular to this strip into a gel containing SDS (Figure 9.5). The final pattern depends on separation according to isoelectric point in one dimension and according to subunit size in the other direction. Crude cell extracts can give up to 1000 individually identifiable protein components in two-dimensional systems. However, for analysis of a purified enzyme it should only be necessary to carry out one-dimensional electrophoresis in each of the two systems. Thus two-dimensional electrophoresis is not a great advantage for purified preparations.

## Staining and Destaining of Proteins in Gels

It is not intended to give full details of methodology in this section, but only to outline possible procedures. After electrophoresis protein zones must be visualized, and this is carried out by a staining procedure, normally involving an organic dye which binds tightly to proteins. Many different dyes can be used, and the main objectives are (i) sensitivity of detecting small amounts of protein and (ii) proportional staining with all types of protein. Most dyes used tend to be attracted by positively charged groups (lysines and arginines) on the proteins; consequently proteins with higher proportions of these—generally more basic proteins—tend to stain more strongly. Indeed, some acidic polypeptides have escaped detection because they bind so little dye.

    The first step is to fix the proteins by denaturation (if not already denatured) in acid. The usual fixatives are either a methanol/acetic acid/water mixture of proportions 3:1:6, or a protein precipitant, trichloracetic acid, or perchloric acid at 5–10% concentration. Except in the cases of sodium dodecyl sulfate gels or isoelectric focusing, the dye may be included in the fixative. For SDS gels the dodecyl sulfate must be washed out first, or the proteins will not bind dye. Ampholytes also bind dye, and it is usually best to allow them to diffuse out during the fixing process, before staining.

Staining is done using the dye dissolved in the fixative. There are many different methods published for overcoming various problems, including the speed of the staining process (132,183–189; these references are by no means exhaustive). Dyes used include Amido Black, Nigrosine, Coomassie Blue G-250, and Coomassie Blue R-250. The latter, R-250, is the most widely used dye for staining proteins. Although it does not stain some acidic proteins as well as, say, Amido Black, Coomassie Blue has generally the greater sensitivity. Some methods (e.g., Ref. 189) depend not on dyes, but on precipitation of potassium or sodium dodecyl sulfate in the cold gel. Contrasting areas contain more or less dodecyl sulfate; protein bands tend to contain less and show up clear against the white background. The method gives a rapid result for SDS gels, but it is not particularly sensitive. Prelabeling of proteins with fluorescamine (190) has also been used successfully. In this case the proteins (denatured, but in SDS gels) can be observed during the electrophoresis by their fluorescence in near-UV light.

Washing out excess stain can be time-consuming. It is necessary to stain long enough for the dye to penetrate to the center of the gel; then excess dye must diffuse out again. The time depends very much on the gel thickness— hence the modern trend toward very thin gels of 1 mm or less. It may be noted here that starch gels, depending only on surface protein, can be stained in a few minutes and washed quickly after that.

The most recent development is more sensitive even than Coomassie Blue; it involves a pseudo-photographic method of silver staining. After fixing the proteins the gel is treated with silver nitrate. The silver binds to proteins (nonstoichiometrically), and after a reductive and enhancement step the process is completed in a few hours (191). This procedure will prove particularly useful when only very limited amounts of sample are available. It also enables a gel to be run with less distortion of bands due to overloading.

The ultimate in sensitivity is to label the sample radioactively with one of the many possible techniques. After fixing (or, more commonly, the gel is frozen to fix it), autoradiography over any desired period can detect very minute amounts.

## Specific Enzyme Staining

One way of determining which of the components found on electrophoresis is the desired enzyme is to use a specific enzyme-staining technique. Applications of this process are clearly limited to electrophoresis in nondenaturing conditions; of the methods described above, urea and SDS gel electrophoresis denature the enzymes. Also, it must be carried out without fixing the protein bands, yet the reagents must be able to diffuse in to the enzyme site before the enzyme itself diffuses in the gel. The most satisfactory systems are those in which the gel is sliced through the protein zones (Figure 9.6), exposing the enzyme on the surface and eliminating the need for diffusion of reagents. Starch or (thick) polyacrylamide slabs are most suited, after running in simple electrophoretic systems, in gradient gels or isoelectric focusing.

There are so many individual methods that only a brief outline of different principles will be given. The gel can be treated by immersion in the reagents, provided that the final product that is visualized (a dye or precipitate) is insoluble, and so stays at the place where it is formed. A soluble color formed at the site of the enzyme will quickly disperse into the bulk liquid. Dehydrogenases are detected by precipitation of *insoluble* reduced tetrazolium dyes, reduced by NAD(P)H being produced at the site of the dehydrogenase.

Alternatively, the reagents may be partially immobilized by gelation (e.g., setting in an agar gel on the surface of the electrophoresis gel) or by soaking into filter paper. In these cases a complex chain of reactions initiated by the presence of the enzyme being detected can be localized; the visible product may be soluble, but cannot diffuse away rapidly. Processes akin to coupled enzymic assays (cf. section 8.3) can be made to occur; production or oxidation of NAD(P)H can be observed under near-ultraviolet light (blue-green fluorescence). Other reaction systems not practical in enzyme assays may be used on gels for qualitative detection. For instance, phosphate liberation by various phosphatases can be trapped as a precipitate of calcium phosphate (192). If the calcium phosphate precipitate is not sufficiently clear, after washing excess reagents away, the calcium phosphate can be converted to lead phosphate, giving black bands at the site of the original enzyme activity.

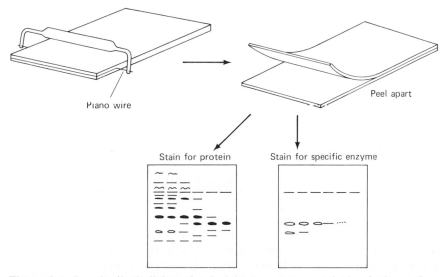

**Figure 9.6.** Longitudinal slicing of gel slabs to expose protein bands for surface staining.

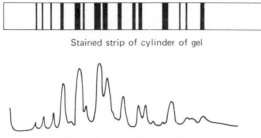

Stained strip of cylinder of gel

Absorption scan of same strip

**Figure 9.7.** Scanning of stained gels for quantitation of protein zones. It is also possible to scan gels at 280 nm and detect unstained proteins by their natural absorption at this wavelength.

It is most important when presenting results of purity from gel electrophoretic analysis that some quantitative estimate of any impurity bands is given. It is not satisfactory to say "gave a single band on electrophoresis," if so little protein was applied that only one faint band showed up. It is also desirable to give some probability that the band being referred to is indeed the enzyme, and not another protein which happens to be present in even greater amounts. There are many examples in the literature where an enzyme has been purified, but not to homogeneity, and the major band on the electrophoretic pattern has been assessed to be the enzyme without any evidence. With very-low-specific-activity final preparations, the actual enzyme may be only 1% of the total protein, and not even noticed as a band on the gel! Presentation of photographs can assist, though relative intensities can rarely be reproduced faithfully. A diagram of a desitometer scan of the stained gel (Figure 9.7) is most convincing. The only reservations then concern the relative *specific* staining intensities of the various components revealed. Making the assumption that all proteins stain equally, precise quantitation of the components can be made with a gel-scanning system.

In summary, analytical gel electrophoresis is one of the most widely used techniques in the biochemical laboratory (and now extensively used for nucleic acids as well as proteins), and characterization of a purified enzyme by one of the techniques described is virtually obligatory. Recent improvements in commercial apparatus design for running thin polyacrylamide gels allow the inexperienced the opportunity of obtaining high-resolution results easily and quickly. The methods are by no means exclusive; analysis on two different principles provides more information on the likely purity and characteristics of the final enzyme preparation.

## 9.2   Other Analytical Methods

Because of the importance and universality of analytical gel electrophoresis, this section describing other methods is very brief; other methods have only limited uses. The object of analyzing the final product is to determine whether or not it consists of one protein or more, and it should be possible to detect quite small levels of impurity. Gel electrophoresis can detect a single impurity component at 1% of the level of the main component (provided that the two are well resolved), but there are occasions when electrophoresis is not suited to the sample. This would particularly apply to lipoproteins and other membrane-associated preparations, which might behave abnormally in electrophoresis and require the presence of certain detergents to maintain structural integrity. In these cases ultracentrifugation may be a better, or at least an additional, method of obtaining information on heterogeneity. Sedimentation velocity experiments will resolve components of quite different $S_{20,w}$ values, but if the impurity is close to the main component, and particularly if it is relatively low in amount, it may not be convincingly demonstrated. Sedimentation equilibrium is more capable of detecting small amounts of impurity through deviation from the theoretical straight line of log (concentration) versus (radius)$^2$ through the sedimentation cell. However, deviations could have interpretations other than the presence of impurities—for instance, a dimerization of the main component.

N-terminal analysis of proteins can give an indication of homogeneity. Provided that the N-terminal amino acid is not acetylated or otherwise blocked, identification of a single type of amino acid with no traces of other minor components is suggestive of a single protein present. However, the conclusions must be qualified by statements indicating that other proteins could be present but would not be detected if (a) they had blocked N-terminals or (b) they had an N-terminal amino acid identical to the main component.

Amino acid composition can also give an idea of the likely heterogeneity of the preparation if two or more amino acid components can be measured very accurately and if the molecular weight of the enzyme subunits is known. If the molar content of these amino acids cannot be expressed by a simple ratio related to the known size of the polypeptide present, then it is probable that a major impurity is confounding the calculation. For instance, if the molecular weight of the polypeptide is known to be close to 30,000 daltons, and tryptophan and tyrosine have been determined spectrophotometrically (193) to be present in amounts of 5.1 and 6.5 residues per 30,000, either these determinations are incorrect or there is an impurity with a different ratio of tryptophan and tyrosine present. For obvious reasons such arguments can only be convincing for small polypeptide sizes, and a clear-cut ratio does not prove that the preparation *is* homogeneous. These calculations are of more value when the preparation is known from other methods to be homogeneous but the exact molecular weight is uncertain. Then an exact molar ratio of two or more amino acids enables the molecular weight to be calculated with some confidence.

Use of crystallization as a criterion of homogeneity is described in the next section; it is by no means unambiguous.

To conclude this brief section, it is clear that nonelectrophoretic methods are not very clear-cut and may be laborious. Nevertheless the measurements described above may have been made anyway, as part of the process of characterizing the enzyme, so if the information is available it can be put to good use.

## 9.3   Crystallization of Enzymes

It is appropriate to conclude this book with a section on enzyme crystallization, as the formation of crystals is often the final stage in purifying and studying an enzyme. Part of this section should strictly be in Chapter 3, since crystallization, as a precipitation method, is a genuine and frequently used method for purifying enzymes. But it is more appropriate to discuss it in general at this point. Crystallization is important in at least four areas:

(1)   As a method of purification
(2)   As a confirmation of homogeneity
(3)   As a method of stable storage
(4)   For determining tertiary structure by diffraction techniques

The mode of obtaining crystals is basically the same for all areas (except perhaps area 4—see below), so they can be discussed all together. Crystals form from supersaturated solutions by aggregation of molecules occurring in an ordered repetitive fashion, rather than random aggregation leading to amorphous precipitates. Crystals in most cases represent a thermodynamically more favorable state than amorphous precipitates—otherwise they would not form—and this is superficially surprising since crystallization clearly represents an increase in order of the protein molecules, and so a decrease in entropy, of the protein. However, the thermodynamics of the *whole system* is much more complex; the disordering of previously bound solvent molecules more than compensates for the ordering of the protein molecules. Simple organic molecules can be crystallized from suitable solvents, and the crystals may develop in a few seconds to be several millimeters in size. But protein crystals cannot develop so rapidly because of the much slower diffusion, collision, and rotation speeds of large molecules; to crystallize, molecules must collide at just the right angle to interact with each other in the appropriate way to form a homogeneous crystal. It may take a large number of unproductive collisions before a correctly lined-up collision takes place. So protein crystals grow slowly, especially if they are to attain more than a microscopically visible size. But micro-crystals can form quite rapidly in some cases, and can be recognized by the characteristic birefringence occurring when agitating a suspension of asymmetric particles little larger than the wavelength of light. Micro-crystals are elongated and needle-like, and create quite attractive appearances in suspension when swirled around.

Two crystallographic features which are often (though not necessarily) related are the morphology of the crystal itself and the geometric nature of the molecular packing within the crystal. The latter is called the space group, and need not concern noncrystallographers. The morphology or shape of the crystals is something which can be observed directly (with the aid of a microscope) and is frequently characteristic for a particular enzyme. Needles of triangular, square, or hexagonal cross-section, diamond or square plates, chunky cubes and truncated octahedrons, hexagonal and other bi-pyramids are all observed. Some enzymes can be persuaded to grow in a variety of different morphologies. In general, the larger crystals must be grown slowly, in very clean containers that give few opportunities for seeding multiple-growth areas, and preferably in a vibration-free building. Large crystals are needed for X-ray diffraction studies, and are grown slowly over a period of weeks or months. For other applications, rapidity of crystal growth is more important than size of crystal.

In enzyme purification, crystallization is often a final step; sometimes it does not result in any increase in specific activity, for the preparation may already have been effectively "pure." To present a final step of crystallization in a purification table indicates the enzyme *is* fairly pure, and also implies that this final crystalline preparation is a suitable form for storage. But crystals can develop in fairly impure protein mixtures. Provided that the impurities are not also in the supersaturated state, aggregation will only occur between molecules of the one protein that is supersaturated, and, if it occurs slowly enough, crystals can develop. The original method for purifying rabbit aldolase (194) involved making a muscle extract 50% saturated in ammonium sulfate and removing the precipitate. One can calculate that aldolase constitutes about 15% of the protein left in the supernatant. On increasing the ammonium sulfate concentration to just 52% saturation, aldolase slowly crystallized out; the presence of the other 85% protein did not prevent crystals from forming. One can conclude from this that the formation of crystals of the enzyme does not prove that the solution from which the crystals formed contained pure enzyme. However, the crystals themselves are not likely to contain much in the way of contaminants, unless there was simultaneous crystallization or amorphous aggregation of an impurity as the enzyme crystals formed. In the latter case it is not so much that the enzyme crystals are impure as that it is not possible to collect those crystals without also collecting the other precipitated protein.

Crystallization as a purification step can be very useful, but recovery of activity may not be satisfactorily high if insufficient time is allowed for completion of crystallization. Rapid formation of micro-crystals is no worse than slow development of large crystals; quick crystallization and recrystallization steps have been useful on a number of occasions (35,195). Successive recrystallizations should reach a steady state of no further improvement in specific activity—another criterion of homogeneity.

Storage of purified proteins is an important benefit of crystallization. Many commercial enzyme preparations are available as crystalline suspensions not just to demonstrate how pure they are, but because such suspensions are the most stable method of storing enzymes (apart from deep-freezing). This is because the molecules are immobilized and so are unlikely to denature because of thermal agitation. Also, proteolytic enzyme contaminants are most unlikely to co-crystallize (though proteases may well adsorb to the surfaces of the crystals), so most molecules are protected from proteolytic degradation. Finally, the usual conditions for crystallization, i.e., high concentrations of ammonium sulfate, are bacteriocidal and stabilize protein molecules.

## Methods for Crystallization

Proteins can be induced to crystallize by creating solvent conditions that result in the protein solution being supersaturated, leading to protein–protein aggregation. Large crystals form best from a solution that is only just supersaturated, in which no amorphous precipitate occurs. However, by removing protein from solution as the crystals form, the remaining solution loses its supersaturation, and no more crystals can form. As a method for purifying the protein, this leads to considerable losses; much of the protein may be left behind in the supernatant. But by continual adjustment of the solvent conditions the solution can be kept in a supersaturated state until nearly all the protein has crystallized. To get to the supersaturated state, any of the precipitants referred to in Chapter 3 can be used. Although ammonium sulfate is employed most frequently, crystals forming in the salting-in (low ionic strength) range can be very satisfactory for diffraction work, though one must be careful about long-term stability because of possible microbial growth. Organic solvents have occasionally been used successfully, but long-term stability requires low-temperature storage. Polyethylene glycol has been very successful in a number of cases where ammonium sulfate was not successful in growing large crystals (196,197). A number of theoretical and practical approaches to the problem of growing large crystals for diffraction have been published (198–201). Generally it requires an extensive study of ranges of conditions, for slight changes in pH or the presence of small amounts of buffer or metal ions can affect the quality of the crystals substantially. The same enzyme from different species may behave quite differently, and the reason why crystallographers always seem to be using enzymes purified from odd sources is that they often find that a particular species' enzyme happens to crystallize much more readily than most others. Hence, sperm whale, lobster, horse, cat, chicken . . . rather than rat, rabbit, and *E. coli.*

Crystallization for the purpose of stabilization and storage does not depend so critically on conditions. Small microscopic crystals are quite suitable. Indeed, large crystals can be a nuisance when trying to get a measured sample from a stock enzyme bottle. The most appropriate way of crystallization for storage is with ammonium sulfate. The enzyme solution is first "salted out" by complete precipitation at a high concentration of ammonium sulfate. The suspension is redissolved in a small volume of suitable buffer, then saturated ammonium sulfate solution added dropwise until a faint permanent turbidity develops. The sample is then quickly centrifuged to remove any specks of dirt, hair, and denatured enzyme, and the supernatant put away in the cold. If after a day or so no signs of crystals or further turbidity develop, some more ammonium sulfate should be added.

For *purification* by crystallization the above process should be followed rather more rapidly; some extra ammonium sulfate can be added immediately after centrifugation, and as soon as crystals appear (usually identifiable by a sheen when shaken) more ammonium sulfate can be added; it is important to recover as much enzyme as possible by leaving little in the solution, though there is always the danger of beginning to precipitate the impurities. A second crystallization will enable trace impurities to be left behind; because their concentration decreases each cycle, a higher level of ammonium sulfate would be needed each time to cause their precipitation, whereas no higher level is needed to crystallize what is wanted. The time needed when using ammonium sulfate for crystallizing (cf. section 7.1) is not a worry because of the stabilizing influence of the salt. But, if crystallization is occurring at low ionic strength, crystal formation should occur as rapidly as possible to avoid the dangers of proteolysis and microbial attack.

It is highly satisfactory to be able to crystallize an enzyme after it has been purified. Not only is it demonstrably pure and in a stable condition, but there is the feeling that one can go no further with the purification; the isolation is at an end, and now characterization of the enzyme protein can commence. These notes on crystallization make a suitable ending to this book, and it is hoped that now many enzyme purifiers will make rapid progress with their projects and will come to agree with the author that protein purification is not just a chore, and no longer a serendipitous act, but a respectable and satisfying science in itself.

# General References

A. H. Gordon (1979), Electrophoresis of proteins in polyacrylamide and starch gels. In *Laboratory techniques in biochemistry and molecular biology* (T. S. Work and R. M. Burdon, eds.), Vol. I, Part I, Elsevier/North-Holland, Amsterdam.
B. Hames and D. Rickwood (1981), *Gel electrophoresis of proteins*. IRL Press, London.

# Appendix A

This section contains tables for calculating amounts of precipitant to add to a solution:

**Table I: Solid Ammonium Sulfate.**
Amount of ammonium sulfate (solid) to be added to a solution already at $S_1$% saturation to take it to $S_2$% saturation.

For 1 liter of solution:

$$g = \frac{533(S_2 - S_1)}{100 - 0.3\,S_2}$$

**Table II: Organic Solvent, or Saturated Ammonium Sulfate.**
Volume of miscible liquid (organic solvent or saturated ammonium sulfate) to be added to a solution already containing $C_1$% v/v liquid, to take it to $C_2$%.

For 1 liter of solution:

$$\text{Volume (ml)} = \frac{10(C_2 - C_1)}{100 - C_2}$$

**Table I.** Ammonium Sulfate, Grams to be Added to 1 Liter

| From S₁% \ To S₂%→ | 5 | 10 | 15 | 20 | 25 | 30 | 35 | 40 | 45 |
|---|---|---|---|---|---|---|---|---|---|
| 0 | 27 | 55 | 84 | 113 | 144 | 176 | 208 | 242 | 277 |
| 5 | | 27 | 56 | 85 | 115 | 146 | 179 | 212 | 246 |
| 10 | | | 28 | 57 | 86 | 117 | 149 | 182 | 216 |
| 15 | | | | 28 | 58 | 88 | 119 | 151 | 185 |
| 20 | | | | | 29 | 59 | 89 | 121 | 154 |
| 25 | | | | | | 29 | 60 | 91 | 123 |
| 30 | | | | | | | 30 | 61 | 92 |
| 35 | | | | | | | | 30 | 62 |
| 40 | | | | | | | | | 31 |
| 45 | | | | | | | | | |
| 50 | | | | | | | | | |

**Table II.** Volume of Miscible Solvent, ml to be Added to 1 Liter

| From C₁% \ To C₂%→ | 5 | 10 | 15 | 20 | 25 | 30 | 35 | 40 | 45 |
|---|---|---|---|---|---|---|---|---|---|
| 0 | 52 | 111 | 176 | 250 | 333 | 428 | 538 | 666 | 818 |
| 5 | | 55 | 117 | 187 | 266 | 357 | 461 | 583 | 727 |
| 10 | | | 58 | 125 | 200 | 285 | 384 | 500 | 636 |
| 15 | | | | 62 | 133 | 214 | 307 | 416 | 545 |
| 20 | | | | | 66 | 142 | 230 | 333 | 454 |
| 25 | | | | | | 71 | 153 | 250 | 363 |
| 30 | | | | | | | 76 | 166 | 272 |
| 35 | | | | | | | | 83 | 181 |
| 40 | | | | | | | | | 90 |
| 45 | | | | | | | | | |
| 50 | | | | | | | | | |

| From S$_1$% | 50 | 55 | 60 | 65 | 70 | 75 | 80 | 85 | 90 | 95 | 100 |
|---|---|---|---|---|---|---|---|---|---|---|---|
| 0 | 314 | 351 | 390 | 430 | 472 | 516 | 561 | 608 | 657 | 708 | 761 |
| 5 | 282 | 319 | 357 | 397 | 439 | 481 | 526 | 572 | 621 | 671 | 723 |
| 10 | 251 | 287 | 325 | 364 | 405 | 447 | 491 | 537 | 584 | 634 | 685 |
| 15 | 219 | 255 | 292 | 331 | 371 | 413 | 456 | 501 | 548 | 596 | 647 |
| 20 | 188 | 223 | 260 | 298 | 337 | 378 | 421 | 465 | 511 | 559 | 609 |
| 25 | 157 | 191 | 227 | 265 | 304 | 344 | 386 | 429 | 475 | 522 | 571 |
| 30 | 126 | 160 | 195 | 232 | 270 | 309 | 351 | 393 | 438 | 485 | 533 |
| 35 | 94 | 128 | 163 | 199 | 236 | 275 | 316 | 358 | 402 | 447 | 495 |
| 40 | 63 | 96 | 130 | 166 | 202 | 241 | 281 | 322 | 365 | 410 | 457 |
| 45 | 31 | 64 | 97 | 132 | 169 | 206 | 245 | 286 | 329 | 373 | 419 |
| 50 | | 32 | 65 | 99 | 135 | 172 | 210 | 250 | 292 | 335 | 381 |
| 55 | | | 33 | 66 | 101 | 138 | 175 | 215 | 256 | 298 | 343 |
| 60 | | | | 33 | 67 | 103 | 140 | 179 | 219 | 261 | 305 |
| 65 | | | | | 34 | 69 | 105 | 143 | 183 | 224 | 266 |
| 70 | | | | | | 34 | 70 | 107 | 146 | 186 | 228 |
| 75 | | | | | | | 35 | 72 | 110 | 149 | 190 |
| 80 | | | | | | | | 36 | 73 | 112 | 152 |
| 85 | | | | | | | | | 37 | 75 | 114 |
| 90 | | | | | | | | | | 37 | 76 |
| 95 | | | | | | | | | | | 38 |

| From C$_1$% | 50 | 55 | 60 | 65 | 70 | 75 | 80 | 85 | 90 | 95 |
|---|---|---|---|---|---|---|---|---|---|---|
| 0 | 1000 | 1222 | 1500 | 1857 | 2333 | 3000 | 4000 | 5666 | 9000 | 19000 |
| 5 | 900 | 1111 | 1375 | 1714 | 2166 | 2800 | 3750 | 5333 | 8500 | 18000 |
| 10 | 800 | 1000 | 1250 | 1571 | 2000 | 2600 | 3500 | 5000 | 8000 | 17000 |
| 15 | 700 | 888 | 1125 | 1428 | 1833 | 2400 | 3250 | 4666 | 7500 | 16000 |
| 20 | 600 | 777 | 1000 | 1285 | 1666 | 2200 | 3000 | 4333 | 7000 | 15000 |
| 25 | 500 | 666 | 875 | 1142 | 1500 | 2000 | 2750 | 4000 | 6500 | 14000 |
| 30 | 400 | 555 | 750 | 1000 | 1333 | 1800 | 2500 | 3666 | 6000 | 13000 |
| 35 | 300 | 444 | 625 | 857 | 1166 | 1600 | 2250 | 3333 | 5500 | 12000 |
| 40 | 200 | 333 | 500 | 714 | 1000 | 1400 | 2000 | 3000 | 5000 | 11000 |
| 45 | 100 | 222 | 375 | 571 | 833 | 1200 | 1750 | 2666 | 4500 | 10000 |
| 50 | | 111 | 250 | 428 | 666 | 1000 | 1500 | 2333 | 4000 | 9000 |
| 55 | | | 125 | 285 | 500 | 800 | 1250 | 2000 | 3500 | 8000 |
| 60 | | | | 142 | 333 | 600 | 1000 | 1666 | 3000 | 7000 |
| 65 | | | | | 166 | 400 | 750 | 1333 | 2500 | 6000 |
| 70 | | | | | | 200 | 500 | 1000 | 2000 | 5000 |
| 75 | | | | | | | 250 | 666 | 1500 | 4000 |
| 80 | | | | | | | | 333 | 1000 | 3000 |
| 85 | | | | | | | | | 500 | 2000 |
| 90 | | | | | | | | | | 1000 |

# Solutions for Measuring Protein Concentration

## (i) Biuret Reaction (202)

Dissolve 1.5 g copper sulfate ($CuSO_4 \cdot 5H_2O$) and 6.0 g sodium potassium tartrate in 500 ml water. Add 300 ml 10% sodium hydroxide, and make up to 1 liter with water. If 1 g potassium iodide is also added, the reagent will keep indefinitely in a plastic container.

### Procedure

To 0.5 ml of unknown containing up to 3 mg of protein, add 2.5 ml reagent. Allow to stand for 20–30 min before reading $A_{540}$.

## (ii) Lowry Reaction (171)

Reagent A  Dissolve 0.5 g copper sulfate ($CuSO_4 \cdot 5H_2O$) and 1 g sodium citrate ($Na_3$ citrate) in 100 ml water. This solution will keep indefinitely, and is more stable than the more common copper-tartate mixture (202).

Reagent B  Dissolve 20 g sodium carbonate ($Na_2CO_3$) and 4 g sodium hydroxide in 1 liter water.

Reagent C  To 50 ml Reagent B, add 1 ml Reagent A.

Reagent D  To 10 ml Folin–Ciocalteau reagent, add 10 ml water.

### Procedure

To 0.5 ml of unknown containing up to 0.5 mg protein, add 2.5 ml Reagent C. Mix and let stand 5–10 min. Then add 0.25 ml Reagent D. Mix well and let stand 20–30 min, read color at a convenient wavelength between 600 and 750 nm. Note that variations in the strength of Folin–Ciocalteau reagent may result in unsatisfactory color development. The dilution of this reagent (or the amount added) may be varied to make the pH of the final mixture after adding Reagents C and D between 10.0 and 10.5. Color development is more rapid at higher pH, but it also fades more quickly. For automated systems where the time of reaction is precisely controlled, rapid color development at pH 11.0 may be more suitable, for which the NaOH concentration in reagent B should be increased by 30–40%.

# (iii)   Coomassie Blue Binding (179, 204)

(a)  100 mg Coomassie Brilliant Blue G-250 is dissolved with vigorous agitation in 50 ml 95% ethanol, then mixed with 100 ml 85% phosphoric acid. The mixture is diluted to 1 liter with water, and filtered to remove undissolved dye. This solution is stable for 1–2 weeks.

(b)  60 mg Coomassie Brilliant Blue G-250 is dissolved in 1 liter 3% perchloric acid, and filtered to remove undissolved material. The absorbance at 465 nm should be between 1.3 and 1.5. This solution is stable indefinitely.

## Procedure

To 1.5 ml of sample containing up to 50 $\mu$g protein, add 1.5 ml Coomassie Blue reagent, and read absorbance at 595 nm after 2–30 min. For dealing with concentrated protein solutions ($>5$ mg ml$^{-1}$), it is convenient to take 1.5 ml water or dilute sodium chloride, add the sample from a microliter syringe, mix well, then add the dye reagent. Concentrated protein solution will precipitate on contact with the dye reagent. Reagent (a) above gives a higher absorbance reading than reagent (b), but is less stable.

# References

1. A. L. Demain (1971), Increasing enzyme production by genetic and environmental manipulations. *Meth. Enzymol. 22,* 86–95.
2. T. M. Roberts and G. D. Lauer (1979), Maximizing gene expression on a plasmid using recombination *in vitro. Meth. Enzymol. 68,* 473–482.
3. K. L. Smiley, A. J. Berry, and C. H. Suelter (1967), An improved purification, crystallization, and some properties of rabbit muscle 5'-adenylic acid deaminase. *J. Biol. Chem. 242,* 2502–2506.
4. L. Van de Berg and D. Rose (1959), Effect of freezing on the pH and composition of sodium and potassium phosphate solution: The reciprocal system $KH_2PO_4$-$Na_2HPO_4$-$H_2O$. *Arch. Biochem. Biophys. 81,* 319–329.
5. Y. Orii and M. Morita (1977), Measurement of the pH of frozen buffer solutions by using pH indicators. *J. Biochem. 81,* 163–168.
6. D. D. Tyler and J. Gonze (1967), The preparation of heart mitochondria from laboratory animals. *Meth. Enzymol. 10,* 75–77.
7. M. Tsuchihashi (1923), Blood catalase. *Biochem. Z. 140,* 63–112.
8. N. R. Lazarus, A. H. Ramel, Y. M. Rustrum, and E. A. Barnard (1966), Yeast hexokinase. I. Preparation of the pure enzyme. *Biochemistry 5,* 4003–4016.
9. W. J. Rutter and J. R. Hunsley (1966), Fructose diphosphate aldolase from yeast. *Meth. Enzymol. 9,* 480–486.
10. S. L. Yun, A. E. Aust, and C. H. Suelter (1976), A revised preparation of yeast *(Saccharomyces cerevisiae)* pyruvate kinase. *J. Biol. Chem. 251,* 124–128.
11. I. Takagahara, Y. Suzuki, T. Fujita, J. Yamauti, K. Fujii, J. Yamashita, and T. Horio (1978), Successive purification of several enzymes having affinities for phosphoric groups of substrates by affinity chromatography on P-cellulose. *J. Biochem. 83,* 585–597.
12. Y. Ito, A. G. Tomasselli, and L. H. Noda (1980), ATP:AMP phosphotransferase from baker's yeast. *Eur. J. Biochem. 105,* 85–92.
13. E. de la Morena, I. Santos, and S. Grisolia (1968), Homogenous crystalline phosphoglycerate phosphomutase of high activity: A simple method for lysis of yeast. *Biochim. Biophys. Acta 151,* 526–528.

14. E. A. Schwinghamer (1980), A method for improved lysis of some gram-negative bacteria. *FEMS Microbiol. Lett. 7,* 157–162.
15. F. M. Clarke and C. J. Masters (1976), Interactions between muscle proteins and glycolytic enzymes. *Int. J. Biochem. 7,* 359–365.
16. L. J. Reed and B. B. Mukkerjee (1969), α-Ketoglutarate dehydrogenase complex from *Escherichia coli. Meth. Enzymol. 13,* 55–61.
17. P. Welch and R. K. Scopes (1981), Rapid purification and crystallisation of yeast phosphofructokinase. *Anal. Biochem. 110,* 154–157.
18. R. K. Scopes (1971), An improved procedure for the isolation of 3-phosphoglycerate kinase from yeast. *Biochem. J. 122,* 89–92.
19. H. S. Penefsky and A. Tzagoloff (1971), Extraction of water-soluble enzymes and proteins from membranes. *Meth. Enzymol. 22,* 204–219.
20. D. Keilin and E. F. Hartree (1947), Activity of the cytochrome system in heart muscle. *Biochem. J. 41,* 500–502.
21. H. S. Penefsky (1979), Preparation of beef heart mitochondrial ATPase. *Meth. Enzymol. 55,* 504–508.
22. L. M. Hjelmeland (1980), A nondenaturing zwitterionic detergent for membrane biochemistry: Design and synthesis. *Proc. Nat. Acad. Sci. 77,* 6368–6370.
23. A. J. Furth (1980), Removing unbound detergent from hydrophobic proteins. *Anal. Biochem. 109,* 207–215.
24. P. A. Srere (1980), The infrastructure of the mitochondrial matrix. *Trends Biochem. Sci. 5,* 120–121.
25. C. Tanford (1980), *The hydrophobic effect: Formation of micelles and biological membranes,* 2nd ed., Wiley-Interscience, New York.
26. W. Melander and C. Horvath (1977), Salt effects on hydrophobic interactions in precipitation and chromatography of proteins: An interpretation of the lyotropic series. *Arch. Biochem. Biophys. 183,* 200–215.
27. R. Czok and T. Bücher (1960), Crystallised enzymes from the myogen of rabbit skeletal muscle. *Advan. Prot. Chem. 15,* 315–415.
28. R. K. Scopes and A. Stoter (1982), All the glycolytic enzymes from one muscle extract. *Meth. Enzymol. 90* (in press).
29. R. K. Scopes (1978), Techniques for protein purification. In *Techniques in the life sciences,* Vol. B101, pp. 1–42, Elsevier/North-Holland, Amsterdam.
30. J. M. Curling (ed.) (1980), *Methods of plasma protein fractionation.* Academic Press, New York.
31. B. A. Askonas (1951), The use of organic solvents at low temperature for the separation of enzymes: Application to aqueous rabbit muscle extract. *Biochem. J. 48,* 42–48.
32. A. Polson, G. M. Potgieter, J. F. Largier, G. E. F. Mears, and F. J. Joubert (1964), The fractionation of protein mixtures by linear polymers of high molecular weight. *Biochim. Biophys. Acta 82,* 463–475.
33. M. Sternberg and D. Hershberger (1974), Separation of proteins with polyacrylic acids. *Biochim. Biophys. Acta 342,* 195–206.
34. S. A. Kuby, L. Noda, and H. A. Lardy (1954), Adenosinetriphosphate-creatine transphosphorylase. I. Isolation of the crystalline enzyme from rabbit muscle. *J. Biol. Chem. 209,* 191–201.
35. R. K. Scopes, K. Griffith-Smith, and D. G. Millar (1981), Rapid purification of yeast alcohol dehydrogenase. *Anal. Biochem. 118,* 284–285.

36. A. Yoshida and S. Watanabe (1972), Human phosphoglycerate kinase. I. Crystallisation and characterisation of normal enzyme. *J. Biol. Chem. 247,* 440–445.
37. E. A. Peterson and H. A. Sober (1956), Chromatography of proteins. I. Cellulose ion-exchange adsorbents. *J. Amer. Chem. Soc. 78,* 751–755.
38. R. K. Scopes (1981), Quantitative studies of ion-exchange and affinity elution chromatography of enzymes. *Anal. Biochem. 114,* 8–18.
39. E. A. Peterson (1970), Cellulosic ion-exchangers. In *Laboratory techniques in biochemistry and molecular biology.* (T. S. Work and E. Work, eds.), Vol. 2, Pt. 2, pp. 223–392, North-Holland, Amsterdam.
40. LKB Produkter AB Publications (1981), *DEAE/CM Trisacryl® M for ion-exchange chromatography.* Bromma, Sweden.
41. Pharmacia Fine Chemicals AB Publications (1980), *Ion-exchange chromatography: Principles and methods.* Uppsala.
42. L. A. A. Sluyterman and O. Elgersma (1978), Chromatofocusing: Isoelectric focusing on ion-exchange columns. I. General principles. *J. Chromatogr. 150,* 17–30.
43. L. A. A. Sluyterman and O. Elgersma (1978), Chromatofocusing: Isoelectric focusing on ion-exchange columns. II. Experimental verification. *J. Chromatogr. 150,* 31–44.
44. L. A. A. Sluyterman and J. Wijdenes (1981), Chromatofocusing. III. The properties of a DEAE-agarose anion exchanger and its suitability for protein separations. *J. Chromatogr. 206,* 429–440.
45. L. A. A. Sluyterman and J. Wijdenes (1981), Chromatofocusing. IV. Properties of an agarose polyethyleneimine ion exchanger and its suitability for protein separations. *J. Chromatogr. 206,* 441–447.
46. Pharmacia Fine Chemicals AB Publications (1981), *Chromatofocusing with Polybuffer^{TM} and PBE^{TM}.* Uppsala.
47. D. C. Watts and B. R. Rabin (1962), A study of the "reactive" sulphydryl groups of adenosine 5′-triphosphate-creatine phosphotransferase. *Biochem. J. 85,* 507–516.
48. E. Algar and R. K. Scopes (1979), Yeast phosphoglycerate kinase: Evidence from affinity elution studies for conformational changes on binding of substrates. *FEBS Lett. 106,* 239–242.
49. A. A. Stewart and R. K. Scopes (1978), Phosphoglycerate kinase from ram testes. *Eur. J. Biochem. 85,* 89–95.
50. S. P. Datta and A. K. Grzybowski (1961), pH and acid-base equilibrium. In *Biochemist's handbook* (C. Long, ed.), pp. 19–58, Spon, London.
51. D. D. Perrin and B. Dempsey (1974), *Buffers for pH and metal ion control.* Chapman & Hall, London.
52. H. A. McKenzie and R. M. C. Dawson (1969), pH and buffers and physiological media. In *Data for biochemical research* (R. M. C. Dawson, D. C. Elliott, W. H. Elliott, and K. M. Jones, eds.), pp. 475–508, Oxford University Press, London.
53. E. A. Barnard (1975), Hexokinase from yeast. *Meth. Enzymol. 42,* 6–20.
54. B. N. Pogell (1962), Enzyme purification by selective elution with substrate from substituted cellulose columns. *Biochem. Biophys. Res. Commun. 7,* 225–230.
55. W. S. Black, A. van Tol, S. Fernando, and B. L. Horecker (1972), Isolation of

a highly active fructose diphosphatase from rabbit muscle: Its subunit structure and activation by monovalent cations. *Arch. Biochem. Biophys. 151,* 576–590.

56. R. K. Scopes (1977), Purification of glycolytic enzymes by using affinity elution chromatography. *Biochem. J. 161,* 253–263.

57. R. K. Scopes (1977), Multiple enzyme purifications from muscle extracts by using affinity elution chromatographic procedures. *Biochem. J. 161,* 265–277.

58. F. von der Haar (1974), Affinity elution principles and applications to purification of aminoacyl-tRNA synthetases. *Meth. Enzymol. 34,* 163–171.

59. W. S. Bennet and T. A. Steitz (1978), Glucose-induced conformational change in yeast hexokinase. *Proc. Nat. Acad. Sci. 75,* 4848–4852.

60. C. A. Pickover, D. B. McKay, M. D. Engelman, and T. A. Steitz (1979), Substrate binding closes the cleft between the domains of yeast phosphoglycerate kinase. *J. Biol. Chem. 254,* 11323–11329.

61. F. von der Haar (1973), Affinity elution as a purification method for aminoacyl-tRNA synthetases. *Eur. J. Biochem. 34,* 84–90.

62. R. Chilla, K. M. Doering, G. F. Domagk, and M. Rippa (1973), A simplified procedure for the isolation of a highly active crystalline glucose-6-phosphate dehydrogenase from *Candida utilis. Arch. Biochem. Biophys. 159,* 235–239.

63. S. S. Qadri and J. S. Easterby (1980), Purification of skeletal muscle hexokinase by affinity elution chromatography. *Anal. Biochem. 105,* 299–303.

64. R. J. Yon (1972), Chromatography of lipophilic proteins on adsorbents containing mixed hydrophobic and ionic groups. *Biochem. J. 126,* 765–767.

65. R. J. Yon and R. J. Simmonds (1975), Protein chromatography on adsorbents with hydrophobic and ionic groups: Some properties of *N*-(3-carboxypropionyl)aminodecyl-Sepharose and its interaction with wheat-germ aspartate transcarbamylase. *Biochem. J. 151,* 281–290.

66. R. J. Yon (1977), Biospecific-elution chromatography with "imphilytes" as stationary phases. *Biochem. J. 161,* 233–237.

67. R. J. Yon (1981), Versatility of mixed-function adsorbents in biospecific protein desorption: Accidental affinity and an improved purification of aspartate transcarbamoylase from wheat germ. *Anal. Biochem. 113,* 219–228.

68. J. Porath and L. Fryklund (1970), Chromatography of proteins on dipolar ion adsorbents. *Nature 226,* 1167–1170.

69. J. Porath and N. Farnstedt (1970), Group fractionation of plasma proteins on dipolar ion exchangers. *J. Chromatogr. 51,* 479–489.

70. L. Grossman and K. Moldave (eds.) (1974), Nucleic acid synthesizing systems, *Meth. Enzymol. 29,* 1–230.

71. J. Turkova (1978), *Affinity chromatography.* Elsevier Scientific, Amsterdam.

72. I. P. Trayer (1978), Affinity chromatography. In *Techniques in the life sciences,* Vol. B102, pp. 1–34, Elsevier/North-Holland, Amsterdam.

73. C. R. Lowe (1977), Affinity chromatography: The current status. *Int. J. Biochem. 8,* 177–181.

74. I. P. Trayer, H. R. Trayer, D. A. P. Small, and R. C. Bottomly (1974), Preparation of adenosine nucleotide derivatives for affinity chromatography. *Biochem. J. 139,* 609–623.

75. P. Cuatrecasas, M. Wilchek, and C. D. Anfinsen (1968), Selective enzyme purification by affinity chromatography. *Proc. Nat. Acad. Sci. 61,* 636–643.

76. S. C. March, I. Parikh, and P. Cuatrecasas (1974), A simplified method for

cyanogen bromide activation of agarose for affinity chromatography. *Anal. Biochem. 60*, 149–152.

77. M. Wilchek, T. Oka, and Y. J. Topper (1975), Structure of a soluble super-active insulin is revealed by the nature of the complex between cyanogen-bromide-activated Sepharose and amines. *Proc. Nat. Acad. Sci. 72*, 1055–1058.

78. L. Sunderberg and J. Porath (1974), Preparation of adsorbents for biospecific affinity chromatography. I. Attachment of group containing ligands to insoluble polymers by means of bifunctional oxiranes. *J. Chromatogr. 90*, 87–98.

79. G. S. Bethell, J. S. Ayers, W. S. Hancock, and M. T. W. Hearn (1979), A novel method of activation of cross-linked agaroses with 1,1'-carbonyldiimidazole which gives a matrix for affinity chromatography devoid of additional charged groups. *J. Biol. Chem. 254*, 2572–2574.

80. K. Nilsson and K. Mosbach (1980), *p*-Toluenesulfonyl chloride as an activating agent of agarose for the preparation of immobilized affinity ligands and proteins. *Eur. J. Biochem. 112*, 397–402.

81. K. Nilsson and K. Mosbach (1981), Immobilisation of enzymes and affinity ligands to various hydroxyl group carrying supports using highly reactive sulphonyl chlorides. *Biochem. Biophys. Res. Commun. 102*, 449–457.

82. C.-Y. Lee and A. F. Chen (1980), Immobilized coenzymes and derivatives. In *Pyridine nucleotide coenzymes* (J. Everse, B. M. Anderson, and K. S. Yon, eds.), Academic Press, New York.

83. P. O'Carra, S. Barry, and T. Griffin (1974), Spacer arms in affinity chromatography: Use of hydrophilic arms to control or eliminate non-biospecific adsorption effects. *FEBS Lett. 43*, 169–175.

84. C. R. Lowe (1977), The synthesis of several 8-substituted derivatives of adenosine 5'-monophosphate to study the effect of the nature of the spacer arm in affinity chromatography. *Eur. J. Biochem. 73*, 265–274.

85. C. L. Wright, A. S. Warsy, M. J. Holroyde, and I. P. Trayer (1978), Purification of the hexokinases by affinity chromatography on Sepharose-*N*-aminoacylglucosamine derivatives. *Biochem. J. 175*, 125–135.

86. J. D. Aplin and L. D. Hall (1980), Sepharose 4 B as a matrix for affinity chromatography: A spin-labelling investigation using nitroxides as model ligands. *Eur. J. Biochem. 110*, 295–309.

87. M. R. A. Morgan, E. George, and P. D. G. Dean (1980), Investigation into the controlling factors of electrophoretic desorption: A widely applicable, nonchaotropic elution technique in affinity chromatography. *Anal. Biochem. 105*, 1–5.

88. R. Haeckel, B. Hess, W. Lauterborn, and K. Wurster (1968), Purification and allosteric properties of yeast pyruvate kinase. *Hoppe-Seyler's Z. Physiol. Chem. 349*, 699–714.

89. G. Kopperschläger, R. Freyer, W. Diezel, and E. Hofmann (1968), Some kinetic and molecular properties of yeast phosphofructokinase. *FEBS Lett. 1*, 137–141.

90. S. T. Thompson, K. H. Cass, and E. Stellwagen (1975), Blue dextran–Sepharose: An affinity column for the dinucleotide fold in proteins. *Proc. Nat. Acad. Sci. 72*, 669–672.

91. J. F. Biellmann, J. P. Samama, C. I. Bränden, and H. Eklund (1979), X-ray studies on the binding of Cibacron Blue F3G-A to liver alcohol dehydrogenase. *Eur. J. Biochem. 102*, 107–110.

92. D. H. Watson, M. J. Harvey, and P. D. G. Dean (1978), The selective retar-

dation of NADP⁺-dependent dehydrogenases by immobilised Procion HE-3B. *Biochem. J. 173*, 591–596.

93. Amicon Corporation Publications (1980), *Dye ligand chromatography: Applications, methods, theory of Mātrex^{TM} gel media.* Lexington, Mass.

94. A. Ashton and G. M. Polya (1978), The specific interaction of Cibacron and related dyes with cyclic nucleotide phosphodiesterase and lactate dehydrogenase. *Biochem. J. 175*, 501–506.

95. Y. D. Clonis and C. R. Lowe (1980), Triazine dyes, a new class of affinity labels for nucleotide-dependent enzymes. *Biochem. J. 191*, 247–251.

96. F. Qadri and P. D. G. Dean (1980), The use of various immobilized triazine affinity dyes for the purification of 6-phosphogluconate dehydrogenase from *Bacillus stearothermophilus*. *Biochem. J. 191*, 53–62.

97. T. Atkinson, P. M. Hammond, R. D. Hartwell, P. Hughes, M. D. Scawen, R. F. Sherwood, D. A. P. Small, C. J. Bruton, M. J. Harvey, and C. R. Lowe (1981), Triazine-dye affinity chromatography. *Biochem. Soc. Trans. 9*, 290–293.

98. C. R. Lowe, M. Hans, N. Spibey, and W. T. Drabble ( 1980), The purification of inosine 5′-monophosphate dehydrogenase from *Escherichia coli* by affinity chromatography on immobilised Procion dyes. *Anal. Biochem. 104*, 23–28.

99. S. W. Kessler (1975), Rapid isolation of antigens from cells with a staphylococcal protein A-antibody adsorbent: Parameters of the interaction of antibody–antigen complexes with protein A. *J. Immunol. 115*, 1617–1624.

100. H. Spielman, R. P. Erikson, and C. J. Epstein (1974), The production of antibodies against mammalian LDH-1. *Anal. Biochem. 59*, 462–467.

101. D. M. Livingstone (1974), Immunoaffinity chromatography of proteins. *Meth. Enzymol. 34*, 723–731.

102. A. L. van Wezel and P. van der Marel (1982), The application of immunoadsorption on immobilized antibodies for large scale concentration and purification of vaccines. In *Affinity chromatography and related techniques* (T. C. J. Gribnau, J. Visser, and R. J. F. Nivard, eds.), pp. 282–292, Elsevier Scientific, Amsterdam.

103. J. Vidal, G. Godbillon, and P. Gadal (1980), Recovery of active, highly purified phosphoenolpyruvate carboxylase from specific immunoadsorbent column. *FEBS Lett. 118*, 31–34.

104. D. Biveau and J. Daussant (1981), Immunoaffinity chromatography of proteins: A gentle and simple desorption procedure. *J. Immunol. Meth. 41*, 387–392.

105. P. Singh, S. D. Lewis, and J. A. Schafer (1979), A support for affinity chromatography that covalently binds amino groups via a cleavable connector arm. *Arch. Biochem. Biophys. 193*, 284–293.

106. G. S. Eisenbarth (1981), Application of monoclonal antibody techniques to biochemical research. *Anal. Biochem. 111*, 1–16.

107. D. S. Secher and D. C. Burke (1980), A monoclonal antibody for large scale purification of human leukocyte interferon. *Nature 285*, 446–450.

108. H. P. Hauri, A. Quaroni, and K. J. Isselbacher (1980), Monoclonal antibodies to sucrase/isomaltase: Probes for the study of postnatal development and biogenesis of the intestinal microvillus membrane. *Proc. Nat. Acad. Sci. 77*, 6629–6633.

109. G. Bernardi, M. G. Giro, and C. Gaillard (1972), Chromatography of polypeptides and proteins on hydroxyapatite column: Some new developments. *Biochim. Biophys. Acta 278*, 409–420.

110. E. Gluekauf and L. Patterson (1974), The adsorption of some proteins on hydroxyapatite and other adsorbents used for chromatographic separations. *Biochim. Biophys. Acta 351*, 57–76.

111. Y. Nakagawa and E. A. Noltmann (1965). Isolation of phosphoglucose isomerase from brewer's yeast. *J. Biol. Chem. 240*, 1877–1881.

112. D. Keilin and F. E. Hartree (1938), Mechanism of the decomposition of hydrogen peroxide by catalase. *Proc. Roy. Soc. B 124*, 387–405.

113. G. Bernardi (1971), Chromatography of proteins on hydroxyapatite. *Meth. Enzymol. 22*, 325–339.

114. LKB-Produkter AB Publications (1980), *HA-Ultrogel®: Hydroxyapatite-agarose gel for adsorption chromatography*. Bromma, Sweden.

115. Z. Er-el, Y. Zaidenzaig, and S. Shaltiel (1972), Hydrocarbon-coated Sepharoses: Use in the purification of glycogen phosphorylase. *Biochem. Biophys. Res. Commun. 49*, 383–390.

116. B. H. J. Hofstee and N. F. Otillio (1978), Modifying factors in hydrophobic protein binding by substituted agaroses. *J. Chromatogr. 161*, 153–163.

117. J. L. Ochoa (1978), Hydrophobic (interaction) chromatography. *Biochimie 60*, 1–15.

118. L. G. Hoffmann (1969), Solubility chromatography of serum proteins. I. Isolation of the first component of complement from guinea pig serum by solubility chromatography at low ionic strength. *J. Chromatogr. 40*, 39–52.

119. L. G. Hoffmann and P. W. McGivern (1969), Solubility chromatography of serum proteins. II. Partial purification of the second component of guinea pig complement by solubility chromatography in concentrated ammonium sulphate solutions. *J. Chromatogr. 40*, 53–61.

120. T. Fujita, Y. Suzuki, J. Yamauti, I. Takagahara, K. Fujii, J. Yamashita, and T. Horio (1980), Chromatography in presence of high concentrations of salts on columns of celluloses with and without ion exchange groups (hydrogen bond chromatography). *J. Biochem. 87*, 89–100.

121. F. von der Haar (1976), Purification of proteins by fractional interfacial salting out on unsubstituted agarose gels. *Biochem. Biophys. Res. Commun. 70*, 1009–1013.

122. F. von der Haar (1978), The ligand-induced solubility shift in salting out chromatography: A new affinity technique, demonstrated with phenylalanyl- and isoleucyl-tRNA synthetase from baker's yeast. *FEBS Lett. 94*, 371–374.

123. L. K. Reed and B. D. Mukherjee (1969), α-Ketoglutarate dehydrogenase complex from *Escherichia coli. Meth. Enzymol. 13*, 55–61.

124. Amicon Corporation Publications (1981), *Boronate ligands in biochemical separations: Applications, method, theory of Mātrex gel PBA*. Lexington, Mass.

125. J. Porath and P. Flodin (1959), Gel filtration: A method for desalting and group separation. *Nature 83*, 1657–1659.

126. J. R. Whitaker (1963), Determination of molecular weights of proteins by gel filtration on Sephadex. *Anal. Chem. 35*, 1950–1953.

127. P. Andrews (1965), The gel filtration behaviour of proteins related to their molecular weight over a wide range. *Biochem. J. 96*, 595–606.

128. O. Smithies (1955), Zone electrophoresis in starch gels: Group variations in the serum proteins of normal human adults. *Biochem. J. 61*, 629–641.

129. L. Ornstein (1964), Disc electrophoresis. I. Background and theory. *Ann. N.Y. Acad. Sci. 121*, 321–349.

130. R. Rüchel and M. D. Brager (1975), Scanning electron microscopic observations of polyacrylamide gels. *Anal. Biochem. 68,* 415–428.

131. C. J. O. R. Morris and P. Morris (1971), Molecular-sieve chromatography and electrophoresis in polyacrylamide gels. *Biochem. J. 124,* 517–528.

132. K. Weber and M. Osborn (1969), The reliability of molecular weight determination by dodecyl sulphate-polyacrylamide gel electrophoresis. *J. Biol. Chem. 244,* 4406–4412.

133. W. F. Goodman and J. N. Baptist (1979), Isoelectric point electrophoresis: A new technique for protein purification. *J. Chromatogr. 179,* 330–332.

134. S. Raymond (1964), Protein purification by elution convection electrophoresis. *Science 146,* 406–407.

135. T. Jovin, A. Chrambach, and M. A. Naughton (1964), An apparatus for preparative temperature-regulated polyacrylamide gel electrophoresis. *Anal. Biochem. 9,* 351–369.

136. C. E. Furlong, C. Cirakoglu, R. C. Willis, and P. A. Santy (1973), A simple preparative polyacrylamide disc gel electrophoresis apparatus: Purification of three branched-chain amino acid binding proteins from *Escherichia coli. Anal. Biochem. 51,* 297–311.

137. J. I. Azuma, N. Kashimura, and T. Komano (1977), A new convenient column for gel electrophoresis, isoelectric focusing, and zonal electrophoresis. *Anal. Biochem. 81,* 454–457.

138. J. A. L. I. Walters and W. S. Bont (1979), An improved method for the separation of proteins by zonal electrophoresis on density gradients. *Anal. Biochem. 93,* 41–45.

139. M. D. Orr, R. L. Blakeley, and D. Panagou (1972), Discontinuous buffer systems for analytical and preparative electrophoresis of enzymes on polyacrylamide gel. *Anal. Biochem. 45,* 68–85.

140. T. C. Bøg-Hansen (1973), Crossed immuno-affinoelectrophoresis: An analytical method to predict the result of affinity chromatography. *Anal. Biochem. 56,* 480–488.

141. V. Hořejší and J. Kocourek (1974), Affinity electrophoresis: Separation of phytohemagglutinins on *O*-glycosyl polyacrylamide gels. *Meth. Enzymol. 34,* 178–181.

142. M. Tichá, V. Hořejší, and J. Barthová (1978), Affinity electrophoresis of proteins interacting with blue dextran. *Biochim. Biophys. Acta 534,* 58–63.

143. V. Hořejší (1981), Review: Affinity electrophoresis. *Anal. Biochem. 112,* 1–8.

144. P. G. Righetti, E. Gianazza, O. Brenna, and E. Galante (1977), Isoelectric focusing as a puzzle. *J. Chromatogr. 137,* 171–181.

145. N. Y. Nguyen, D. Rodbard, P. J. Svendsen, and A. Chrambach (1977), Cascade stacking and cascade electrofocusing: Their interconversion and fundamental unity. *Anal. Biochem. 77,* 39–55.

146. N. Y. Nguyen and A. Chrambach (1977), Natural pH gradients in buffer mixtures: Formation in the absence of strongly acidic and basic anolyte and catholyte, gradient steepening by sucrose, and stabilization by high buffer concentrations in the electrolyte chambers. *Anal. Biochem. 79,* 462–469.

147. R. L. Prestidge and M. T. W. Hearn (1979), Preparative flatbed electrofocusing in granulated gels with natural pH gradients generated from simple buffers. *Anal. Biochem. 97,* 95–102.

148. G. Baumann and A. Chrambach (1976), Gram-preparative protein fractionation

by isotachophoresis: Isolation of human growth hormone isohormones. *Proc. Nat. Acad. Sci. 73*, 732–736.

149. A. Dobry and F. Boyer-Kawenoki (1947), Phase separation in polymer solution. *J. Polym. Sci. 2*, 90–100.

150. P.-Å. Albertsson (1971), In *Partition of cell particles and macromolecules,* 2nd ed., pp. 24–25, Wiley, New York.

151. A. Hartman, G. Johansson, and P-Å. Albertsson (1974), Partition of proteins in a three-phase system. *Eur. J. Biochem. 46*, 75–81.

152. A. Chaabouni and E. Dellacherie (1979), Affinity partition of proteins in aqueous two-phase systems containing polyoxyethylene glycol-bound ligand and charged dextrans. *J. Chromatogr. 171*, 135–143.

153. M.-R. Kula (1979), Extraction and purification of enzymes using aqueous 2-phase systems. In *Applied biochemistry and bioengineering* (L.B. Wingard, E. Katchalski-Katzir and L. Goldstein, eds.), Vol. 2, pp. 71–96, Academic Press, New York.

154. W. Müller, H. Bünemann, H.-J. Schuetz, and A. Eigel (1982), Nucleic acid interacting dyes suitable for affinity chromatography, partitioning, and affinity electrophoresis. In *Affinity chromatography and related techniques* (T. C. J. Gribnau, J. Visser, and R. J. F. Nivard, eds.), pp. 437–444, Elsevier Scientific, Amsterdam.

155. J. A. Illingworth (1981), A common source of error in pH measurements. *Biochem. J. 195*, 259–262.

156. W. W. Cleland (1964), Dithiothreitol, a new protective reagent for SH groups. *Biochemistry 3*, 480–482.

157. C. de Duve and P. Baudhuin (1966), Peroxisomes (microbodies and related particles). *Physiol. Rev. 46*, 323–357.

158. Nomenclature Committee of the International Union of Biochemistry (1978), *Enzyme nomenclature.* Academic Press, New York.

159. A. M. Gold (1967), Sulfonylation with sulfonyl halides. *Meth. Enzymol. 11*, 706–711.

160. H. Umezawa, T. Aoyagi, H. Morishima, M. Matsuzak, M. Hamada, and T. Takeuchi (1970), Pepstatin: A new pepsin inhibitor produced by actinomyces. *J. Antibiot. 23*, 259–262.

161. I. Kregar, I. Uhr, R. Smith, H. Umezawa, and V. Turk (1977), In *Intracellular protein catabolism* (V. Turk and N. Marks, eds.), Vol. 2, pp. 250–254, Plenum Press, New York.

162. K. Gekko and S. N. Timasheff (1981), Thermodynamic and kinetic examination of protein stabilization by glycerol. *Biochemistry 20*, 4677–4686.

163. M. B. Smith, D. G. Oakenfall, and J. F. Back (1978), Thermal stabilization of protein by sugars and polyols. *Proc. Aust. Biochem. Soc. 11*, 4.

164. G. Ramponi, A. Guerritore, C. Treves, P. Nassi, and V. Baccari (1969), Horse muscle acyl phosphatase: Purification and some properties. *Arch. Biochem. Biophys. 130*, 362–369.

165. A. E. Aust and C. H. Suelter (1978), Homogenate pyruvate kinase isolated from yeast by two different methods is indistinguishable from pyruvate kinase in cell-free extract. *J. Biol. Chem. 253*, 7508–7512.

166. F. C. Wormack and S. P. Colowick (1979), Proton-dependent inhibition of yeast and brain hexokinases by aluminum in ATP preparations. *Proc. Nat. Acad. Sci. 76*, 5080–5084.

167. B. Crabtree, A. R. Leech, and E. A. Newholme (1979), Measurement of enzymic activities in crude extracts of tissues. In *Techniques in the life sciences,* Vol. B211, pp. 1–37, Elsevier/North-Holland, Amsterdam.

168. R. K. Scopes (1972), Automated fluorometric analysis of biological compounds. *Anal. Biochem. 49,* 73–87.

169. F. Garciá-Carmona, F. Garciá-Cánaves, and J. A. Lozano (1981), Optimizing enzyme assays with one or two coupling enzymes. *Anal. Biochem. 113,* 286–291.

170. R. F. Itzhaki and D. M. Gill (1964), A micro-biuret method for estimating protein. *Anal. Biochem. 9,* 401–410.

171. O. H. Lowry, N. J. Rosebrough, A. L. Farr, and R. J. Randall (1951), Protein measurement with the Folin phenol reagent. *J. Biol. Chem. 193,* 265–275.

172. O. Folin and V. Ciocalteau (1927), Tyrosine and tryptophan determination in proteins. *J. Biol. Chem. 73,* 627–650.

173. G. L. Peterson (1979), Review of the Folin phenol protein quantitation method of Lowry, Rosebrough, Farr, and Randall. *Anal. Biochem. 100,* 201–220.

174. O. Warburg and W. Christian (1941), Isolierung und Kristallisation des Gärungsferments Enolase. *Biochem. Z. 310,* 384–421.

175. A. R. Goldfarb, L. J. Saidel, and E. Mosovich (1951), The ultraviolet absorption spectra of proteins. *J. Biol. Chem. 193,* 397–404.

176. D. R. Wetlaufer (1962), Ultraviolet spectra of proteins and amino acids. *Advan. Prot. Chem. 17,* 303–390.

177. R. K. Scopes (1974), Measurement of protein by spectrophotometry at 205 nm. *Anal. Biochem. 59,* 277–282.

178. M. M. Bradford (1976), A rapid and sensitive method for the quantitation of microgram quantities of protein utilizing the principle of protein-dye binding. *Anal. Biochem. 72,* 248–254.

179. T. Spector (1978), Refinement of the Coomassie blue method of protein quantitation. *Anal. Biochem. 86,* 142–146.

180. J. A. Reynolds and C. Tanford (1970), Binding of dodecyl sulphate to proteins at high binding ratios: Possible implications for the state of proteins in biological membranes. *Proc. Nat. Acad. Sci. 66,* 1002–1007.

181. S. Cavell and R. K. Scopes (1976), Isolation and characterisation of the "photosynthetic" phosphoglycerate kinase from *Beta vulgaris. Eur. J. Biochem. 63,* 483–490.

182. A. C. Arcus (1970), Protein analysis by electrophoretic molecular sieving in a gel of graded porosity. *Anal. Biochem. 37,* 53–63.

183. R. W. Blakesly and J. A. Boezi (1977), A new staining technique for proteins in polyacrylamide gels using Coomassie Brilliant Blue G250. *Anal. Biochem. 82,* 580–582.

184. W. Diezel, G. Kopperschläger, and E. Hofmann (1972), An improved procedure for protein staining in polyacrylamide gels with a new type of Coomassie Brilliant Blue. *Anal. Biochem. 48,* 617–620.

185. N. Malik and A. Berrie (1972), New stain fixative for proteins separated by gel isoelectric focusing based on Coomassie Brilliant Blue. *Anal. Biochem. 49,* 173–176.

186. M. Ortega (1957), Use of nigrosine for staining proteins after electrophoresis on filter paper. *Nature, 179,* 1086–1087.

187. W. L. Ragland, J. L. Pace, and D. L. Kemper (1974), Fluorometric scanning of

fluorescamine-labelled proteins in polyacrylamide gels. *Anal. Biochem. 59*, 24–33.

188. C. M. Wilson (1979), Studies and critique of amido black 10B, Coomassie Blue R, and fast green FCF as stains for proteins after polyacrylamide gel electrophoresis. *Anal. Biochem. 96*, 263–278.

189. R. C. Higgins and M. E. Dahmus (1979), Rapid visualisation of protein bands in preparative SDS-polyacrylamide gels. *Anal. Biochem. 93*, 257–260.

190. J. L. Pace, D. L. Kemper, and W. L. Ragland (1974), The relationship of molecular weight to electrophoretic mobility of fluorescamine-labeled proteins in polyacrylamide gels. *Biochem. Biophys. Res. Commun. 57*, 482–487.

191. D. W. Sammons, L. D. Adams, and E. E. Nishizawa (1981), Ultrasensitive silver-based colour staining of polypeptides in polyacrylamide gels. *Electrophoresis 2*, 135–140.

192. H. J. Schäfer, P. Scheurich, and G. R. Rathgeber (1978), A simple method of activity staining for phosphate-releasing enzymes in electrophoresis gels. *Hoppe-Seyler's Z. Physiol. Chem. 359*, 1441–1442.

193. H. Edelhoch (1967), Spectroscopic determination of tryptophan and tyrosine in proteins. *Biochemistry 6*, 1948–1954.

194. J. F. Taylor, A. A. Green, and G. T. Cori (1948), Crystalline aldolase. *J. Biol. Chem. 173*, 591–604.

195. R. K. Scopes (1974), A simple procedure for crystallisation of marsupial muscle phosphorylase. *Proc. Aust. Biochem. Soc. 7*, 6.

196. A. McPherson (1976), Crystallisation of proteins from polyethylene glycol. *J. Biol. Chem. 251*, 6300–6306.

197. T. Albert, F. C. Hartman, R. M. Johnson, G. A. Petsko, and D. Tsernoglou (1981), Crystallisation of yeast triose phosphate isomerase from polyethylene glycol: Protein crystal formation following phase separation. *J. Biol. Chem. 256*, 1356–1361.

198. Z. Kam, H. B. Shore, and G. Faher (1978), On the crystallisation of proteins. *J. Mol. Biol. 123*, 539–555.

199. C. W. Carter, Jr., and C. W. Carter (1979), Protein crystallisation using incomplete factorial experiments. *J. Biol. Chem. 254*, 12219–12223.

200. B. H. Weber and P. E. Goodkin (1970), A modified microdiffusion procedure for the growth of single protein crystals by concentration-gradient equilibrium dialysis. *Arch. Biochem. Biophys. 141*, 489–498.

201. R. Henderson (1980), Crystallising membrane proteins. *Nature 287*, 490.

202. A. C. Gornall, C. J. Bardawill, and M. M. David (1949), Determination of serum proteins by means of the biuret reaction. *J. Biol. Chem. 177*, 751–766.

203. J. Leggett-Bailey (1967), *Techniques in protein chemistry*, 2nd ed., pp. 340–346, Elsevier, Amsterdam.

204. J. J. Sedmak and S. E. Grossberg (1977), A rapid, sensitive and versatile assay for protein using Coomassie Brilliant Blue G250. *Anal. Biochem. 79*, 544–552.

# Index